AF773786

B&M/BRIZIO Project Roadster

ROY BRIZIO STREET RODS

MODERN HOT RODS DEFINED

BO BERTILSSON

motorbooks

First published in 2009 by Motorbooks, an imprint of MBI Publishing Company, 400 First Avenue North, Suite 300, Minneapolis, MN 55401 USA

ISBN-13: 978-0-7603-3544-4

Library of Congress Cataloging-in-Publication Data

Bertilsson, Bo, 1946-
Roy Brizio Street Rods : modern hot rods defined /
Bo Bertilsson ;
foreword by Vic Edelbrock.
p. cm.
ISBN 978-0-7603-3544-4 (hb w/ jkt)
1. Hot rods--United States--History.
2. Automobiles--Customizing--United States--History. 3. Brizio, Roy. 4. Roy Brizio Street Rods (Firm) 5. Hot rods--United States--Pictorial
works. I. Edelbrock, Vic, 1936- II. Title.

TL236.3.B4733 2009
629.228'60973--dc22

2009001309

On the cover: Cliff Hanson's '31 Roadster Pickup. See page 58.

On the frontispiece: Steve Coonan snaps a shot while driving Roy's blue and red scalloped B&M/Brizio project roadster through the desert. *Steve Coonan*

On the title pages: *(top left)* Jeff Beck's Boogie-Woogie three-window coupe. See page 69. *(top right)* Jorge Zaragoza's Golden '36 Coupe. See page 127. *(bottom)* Roy Brizio's '55 Chevy, *The Golden Shoebox*. See page 144.

On the back cover, main: *(left)* Reggie Jackson and Roy Brizio, circa 1980. Roy Brizio Collection. *(right)* Roy Brizio and his purple '32 roadster.

Editors: Lindsay Hitch and Jeffrey Zuehlke
Design: John Sticha
Jacket Design: John Barnett

Printed in China

CONTENTS

CHAPTERS

ACKNOWLEDGMENTS

Among the people that made this book possible, I wish to thank my friend, Roy Brizio, and his family for all their time and help. Also, Steve Coonan and Geoff Miles for helping out with pictures and letting me use the *Rodder's Journal* studio; and Jorge Zaragoza for letting me photograph his car collection.

Bo Bertilsson

FOREWORD
BY VIC EDELBROCK

MY DEFINITION OF A STREET ROD

America has a love affair with the automobile. Much of this romance has to do with what we call "street rods": cars from the model years before 1949 or cars leading up to the muscle cars of the 1960s.

A street rod is the car of your dreams. It grabs your heart and fuels your passion when you get in the seat and drive it for the first time. It becomes part of the family. Your family can enjoy the performance and beauty of the car. Your dream has become a reality.

This is where Roy Brizio comes into the picture. Call Roy; he and his company have become the dream makers. I first met Roy at the 1979 SEMA show through an introduction from baseball legend Reggie Jackson. Roy had just finished a '32 roadster for Reggie, and it was displayed in the Accell SEMA booth. Seeing Reggie's '32 brought back memories of Vic Sr. and my mother Katie's first everyday car. It was a car they drove during the week, and my dad raced it at the dry lakes one weekend a month during spring, summer, and early fall. Vic Sr.'s real '32 was owned by Jim Ladley of Santa Rosa, and it was not for sale. So I contacted Roy to build an exact replica and flew to South San Francisco, along with Bobby Meeks, who had been friends with my dad since I was born in 1936. He knew the original car backward and forward, and using scrapbooks and many pictures for reference, we started the project. What a great project it was, indeed. It was my first dream car, and it started my longtime relationship with Roy. My dream came true. The car was delivered to the SEMA show and was displayed in our booth.

At the 1986 SEMA show I saw a red highboy '32 with a Ferrari engine that Roy had built for one of his customers. After that, I started to dream about car number two. Then Roy built me a beautiful car once again. (In 2006, I had Roy re-do the car to match what we are doing today, and it was displayed at SEMA in our booth in 2007.)

Every time we drove Roy's first two cars, my wife, Nancy, would tell me that she wanted a 1946 woody station wagon. I gave in without her knowing it and asked Roy to look into finding a '46 woody. He called me a week later and told me he found a car. I sent a check, and off Roy went with a finish date to match the Christmas party for my employees, where my father's Midget No. 27 V8-60 was being introduced after restoration. Nancy was completely surprised when we opened the car trailer and out came a beautiful '46 woody. Roy took care of everything. What a dream come true—a complete surprise—for Nancy. She had no idea. (A side note: Roy sent Nancy a cookie jar with a woody on it for Christmas a year before. I told Nancy at the time that was as close as she was going to get to owning a woody.)

Next dream car. When I was 16 years old, I wanted a '40 Ford coupe for my first car. I couldn't find anyone who wanted to sell, so I settled for a '46 Ford convertible, which I had for five years. (I never should have sold it.) Each year at SEMA, we would have a party in our Edelbrock suite on Wednesday night. It was a great party, and a few cocktails were consumed. On one occasion, Roy and a friend from San Francisco, along with Nancy and me, were the last ones still standing. Our conversation, of course, was about cars, and somehow it got on '40 Fords. Roy's friend said he had a very good steel body, and another dream car was started. It was delivered out of the box for the Hot Rod Power Tour. (It was flamed without my knowledge, because I was against it, but I have to admit it really made the car look great.) The '40 Ford is a California car and hadn't been exposed to rain, but we started with the car in Kansas City, where they were having floods and lots of rain. Nancy was in charge of the situation, using many towels from the motels to keep us dry.

The car turned out great, though we never let Roy forget about flood control.

Next came the delivery '40 Ford sedan. My dad did not have one, although there is a picture of my dad's shop on Highland in Hollywood with a 1940 delivery van parked in front. That van belonged to Harman and Collins, and I asked Roy to replicate it. We designed the old Edelbrock logo "with established in 1938 on it."

On to the next project. Jim Ladley had my dad's car, and we worked out a deal, but that deal would happen only after Jim died. Not knowing when that would eventually happen and knowing that I needed the services of Bobby Meeks (who knew how my dad's car was really built), I called Roy to see if he could find another real '32 roadster. He did, and he made the car just as it was back in the day: a daily driver with fenders. It really came out great and was in our booth at SEMA—another dream car.

For dream car number seven, Roy called me regarding a project for a TLC cable television show. The scenario was that I would buy a car from Roy and the car would be built on television. I said, "Go for it." The producers chose a '32, and it became a great car for a show. The car was for our daughters, Camee and Christi, and they picked the colors and flaming. I asked Camee to show me the sample colors and said I would like to see samples with my color red. Camee said, "Sorry, Dad, the car is already painted." And I said, "Why are you showing me the sample, then?" I got even, though, since the license plate says "VICS-32." This car has a great following and was displayed in our booth at SEMA.

My number eight dream car is Vic Sr. and Katie's original '32. After Jim Ladley passed away, the family called Roy and asked if we still wanted the car. Roy went to their house in Santa Rosa that same afternoon to pick up the car. It had been changed over the years, and Roy restored it to original condition. We made it just as it looked on the dry lakes at Muroc three weeks before Pearl Harbor in 1941, when Vic Sr. went almost 122 miles per hour. That was the high point for the car. The car finally came home after being displayed at SEMA.

Thank you, Roy, for making my car dreams come true. You have been a very close friend of the family for 30 years and will be for many more years to come. You have also made your customers' dreams come true.

Great job, my friend.

—Vic Edelbrock

Looking back at some of the magazine covers that have featured Brizio cars through the years, we picked 50 for this shot.

PREFACE

Working in the magazine business for the last 35 years and photographing mostly hot rods and customs, I have been in contact with Roy Brizio many, many times. Roy has always given me a chance to take pictures of his latest cars as soon as they were finished. With more than three hundred cars built by his shop through the years, I've been able to feature many of Brizio's cars in European magazines

English rock guitar player Jeff Beck is also a hot rodder with a garage full of '32 Fords. In 1975, while working for the magazine *Colorod* in my home country of Sweden, I took the ferry to see Jeff in London. I had heard from my friend Tony Thacker that this ex-Yardbirds guitar player had some neat cars, and Tony introduced me to Jeff. At the time, Jeff already had a few '32s in his garage, plus an *Andy's Instant T* T-bucket that he built from a kit he bought from Andy Brizio. Jeff did a lot of the work on his own cars rather than hiring a local shop to do it for him. He said it was relaxing to work on the hot rods and a welcome break from recording and touring all over the world. I did a story about Jeff, and it ran in *Colorod* that year. Jeff has been a Brizio customer since the day Roy opened his shop in South San Francisco, and the two men have become good friends. Today, Jeff has eight or nine '32s, and one of them is a three-window coupe that Roy built for him a few years ago, which is included in this book. Through the years, Roy has also built chassis for some of Jeff's other hot rods. A few years after I met Jeff in England, I made my first trip to Roy Brizio's shop in San Francisco. Since then, I've made a lot of trips to the Brizio shop, where they produce more great-looking driver hot rods than any other shop in the business.

Another person who was part of both Roy's and my profession was Ed "Big Daddy" Roth, whom I visited during my first trip over the ocean in 1969. He told me to buy a good camera and start taking pictures for his *Choppers* magazine, which got me started in the magazine business. Roy was seven years old when he first met Ed at the 1963 Grand National Roadster Show in Oakland, and "Big Daddy" made a big impression on him. We each saw Ed periodically through the years, and I know I always learned something when I spent time with the legendary builder. Ed got me on the right track early, and I'm thankful for that today. He made it possible for me to make my hobby a full-time job, which is not easy for many people to do.

Eric Clapton, another famous English rock guitar player, is a customer of Roy's. Eric and Roy were introduced by Jimmie Vaughan, a car-enthusiast rock-and-roller. Together, they checked out the Grand National Roadster Show, and Eric set his sights on a '40 Ford coupe after looking at all the street rods. He asked Roy to find a good one and build it for him. Roy loaned one of his roadsters to Eric for a few weeks to give him a better feel for street rodding. Eric loved the experience and said he was amazed that "nobody saw me; they just checked out the car." Eric liked the '32 roadster so much that he had Roy build him a full-fendered version painted light blue. Next, Roy built him a nice '32 Ford Victoria. A few years later, Eric contacted Roy again to plan some new projects. Right now, Roy Brizio Street Rods is working on a chopped '49 Ford coupe with a modern chassis and drivetrain, and a '49 Chevy pickup. Both projects are built for right-hand steering because Eric will take them back to England.

Needless to say, it has been a lot of fun for me to work on this book. I hope you will enjoy it as well.

1 THE EARLY YEARS

This '32 three-window coupe, shown at Roy Brizio's second shop in San Bruno, belonged to one of his customers.
Roy Brizio Collection

WHEN ROY BRIZIO was just 15 months old, his dad Andy sat him on the cowl of his '32 coupe for a family photo at the 1957 Oakland Roadster Show. Roy grew up with hot rods around him all the time. His dad worked at Western Tires in San Francisco, and on the weekends Andy Brizio was the starter at the Half Moon Bay drag strip. A few years later, Andy opened Andy's Wheels & Tires next door to his drag racing pals at Champion Speed Shop.

By the mid-1960s, Andy became interested in the T-bucket hot rods that were so popular in the late 1950s and early 1960s. When Dragmasters, a chassis-building company in Carlsbad, California, introduced a torsion bar tube chassis for the T-bucket, Andy Brizio became a Dragmasters dealer for northern California. A year later, Andy began making his own chassis in rectangular tubing with the help of local fabricator Pete Ogden. Andy marketed a T-bucket kit called *Andy's Instant T*, which became well known in the hot rod world and appeared in numerous ads and magazine features. Andy drove his *Instant T* to hot rod events in northern California and joined some cross-country tours with *Rod & Custom* magazine. He sold the kits overseas, and one of his British customers was rock guitarist Jeff Beck, who later became Roy Brizio's

Above: At one year and three months old, Roy sits on the cowl of dad's '32 five-window coupe at the 1957 Grand National Roadster Show in Oakland. Roy has been around street rods all his life.
Roy Brizio Collection

Left: Seven-year-old Roy poses with one of his heroes, Ed "Big Daddy" Roth, at the 1963 Oakland Roadster Show.
Roy Brizio Collection

Already a hot rodder at age 11, Roy learned to drive in his dad's original *Instant T*, pictured at a hot rod picnic in 1968. *Roy Brizio Collection*

Roy and his friend Dean Icardi sit in a blown T-bucket at the Roadster Roundup in Pismo Beach in 1967. *Roy Brizio Collection*

Andy Brizio and his best friend, Cub Barnett, bought the Champion Speed Shop in 1972, and Andy ran his Instant T shop next door. Roy first learned to weld and work on hot rods here. *Roy Brizio Collection*

good friend and regular customer. Jeff Beck built his own *Instant T* in his garage in the early 1970s, alongside a few '32 Fords.

Roy rode with his dad to many events in the T-bucket and got very familiar with hot rods at an early age. In 1967, at age 11, he drove his dad's *Instant T* at the Pismo Beach roadster event.

In 1971, 15-year-old Roy went to Los Angeles to spend the summer with family friend Dan Woods, who owned a hot rod shop with Jim "Jake" Jacobs. At the time, the shop was busy building show cars, including the *Roach Coach* for Roach Studios and the *Butcher Truck.* Roy helped at the shop that summer and learned a lot about car building in a short period of time. During the school year, Roy spent as much time at his dad's shop as he could, learning how to weld and fabricate brackets, and how to put hot rods together. Roy got one heck of a hot rod–building education from his dad, the boys in the shop, and his dad's friends in Los Angeles.

The '32 three-window coupe that Roy bought for $1,500 in 1973 and paid off with a bank loan became very important later, when it was parted out and Roy used the money to start his own shop. Boyd Coddington bought the last piece, the body, in 1979. *Roy Brizio Collection*

Reggie Jackson hired Roy to build this '32 roadster for the 1979 SEMA show. With only three months to finish the project, Roy located a car with all the chassis work done, stripped its paint, repainted it, and detailed it with a blown small-block. *Roy Brizio Collection*

In 1972, Andy Brizio and his friend Cub Barnett bought Champion Speed Shop, and they ran both businesses for a few years. The T-bucket trend was strong for about 10 years, and when the orders started to slow down, Andy closed the chassis-building part of the business. He tried to get his son interested in working in the speed shop, but Roy was too interested in hot rods and, most of all, '32 Fords. Sensing it would be hard to keep a hot rod shop afloat during the 1970s, Andy tried to talk Roy out it. But rather than work the counter at the speed shop, Roy focused on helping his dad's *Instant T* customers with repairs and small modifications.

Working on the hot rods led to lots of good contacts, and Roy often got tips on good street rods for sale. In 1973, Roy got a bank loan to buy a '32 three-window for $1,500 from one of his dad's friends. He made $60 monthly payments—a good-sized chunk of change at the time. Later, Roy parted with the coupe, using the money to start his own shop. He sold the blown motor first, and later he sold the fenders for $1,000. In 1979,

The second car Roy and his crew built for Reggie Jackson and Accel Ignition was this nonchopped '32 five-window coupe equipped with a Weber carburetor and a small-block Chevy engine. *Roy Brizio Collection*

In 1980, after Roy and his crew moved to the bigger shop on Wattis Way in South San Francisco, they built two 1932 stake-bed trucks—one for the shop and one for Dennis Varni. Both trucks had small-block Chevys. *Roy Brizio Collection*

Tom Medley, who worked for *Hot Rod* and *Rod & Custom* magazines, added a cartoon hot rodder to this copy of Roy Brizio Street Rods' first catalog, made in 1983. *Roy Brizio Collection*

Roy works on the Wescott fiberglass-bodied roadster that would become a Ferrari engine–powered winner at the 1987 Oakland Roadster Show. This picture was taken at the Wattis Way shop in 1985. *Roy Brizio Collection*

Roy took the body off the chassis and sold it to Boyd Coddington. Roy also owned the C&T Automotive '32 roadster that Andy bought when his son was 14 years old. In 1975, Roy sold that car to Don Thealan, who turned that car into an Oakland Roadster Show winner for wheel manufacturer John Sironian.

By that time, Roy had his own 2,500-square-foot shop in San Bruno and had hired a part-time fabricator. And Andy had sold the speed shop, starting a T-shirt business in 1979. About the same time, race car chassis builder and family friend "Little" John Buttera took Roy to see Roy Fjastad at SPE in Santa Ana. Fjastad had made a set of stamped replica '32 frame rails, which Roy Brizio realized would make quite an impact on the street rod world. During the two-week stay in Los Angeles, Little John also helped Roy make the fixtures for a frame jig.

Back in his San Francisco shop, Roy started to build a chassis jig for the '32 Ford frame. He bought frame rails from Fjastad, who sold his new product under

Roy and the crew—from left: Jack Stratton, Jim Vickery, Roy, and Scott Amaroso—with the James Ellis–owned 1987 Grand National Roadster Show winner. *Roy Brizio Collection*

Through the guys at the Champion Speed Shop, Roy got involved in nostalgia drag racing in the 1980s. This picture of Roy behind the wheel of the Champion Speed Shop blown small-block dragster was taken at the Fremont Drag Strip in 1986. *Roy Brizio Collection*

Andy Brizio arranged Andy's Picnic hot rod event for many years, and this picture was taken with Roy and Andy in 1991. *Roy Brizio Collection*

the Deuce Factory name. The '32 frames generated a lot of business for Brizio, and Roy realized he needed more space. In January 1980, he moved into a 5,000-square-foot shop on Wattis Way, near the San Francisco airport.

Just before the move to his new shop, Roy was introduced to a guy named Reggie Jackson, whom Roy had heard of, although he was unaware that Reggie was a famous baseball player. Reggie needed a car built for the SEMA show, which was only three months away. It would have been impossible for the Brizio crew to build a car that quickly, but as luck would have it, Roy was able to use a project they had started for someone else, a nearly finished '32 roadster with a Wescott fiberglass body. Roy bought the car, stripped it and repainted it yellow with an Accel sponsor logo, and installed a blown small-block between the frame rails. Roy and Reggie loaded the car on a ramp truck and drove it to Las Vegas for the 1979 SEMA show. It was Roy's first trip to the show, which was not well known at the time, even among hot rod builders.

The Wescott fiberglass body was the best, and Wescott's roadster bodies have steel reinforcement.

Roy and Andy Brizio pose with the legendary George Barris (center) at the 50th Autorama in Detroit. *Roy Brizio Collection*

Roy was impressed with the body and ordered five more. He became a Wescott dealer, which helped him sell more bodies and complete cars. The combination of Deuce Factory frame rails, Brizio frames, and Wescott bodies enabled Roy Brizio Street Rods to build numerous roadsters during the 1980s and 1990s. Roy made his first catalog in 1983, listing chassis details, stainless-steel parts, frames, and more. The company kept growing, and Brizio-built cars began to appear in magazines more and more often. Roy's cars have always had a reputation for being well-built, good-looking drivers, something proven regularly by the fact that his customers can—and do—drive them to hot rod events across the nation.

When Brookville Roadsters introduced a replica steel '32 roadster body, the fiberglass roadster bodies became nearly obsolete. Since that time, the Brizio shop has always tried to use original Ford steel bodies whenever possible. The latest Brookville replica—the '32 three-window coupe body—is identical to the original Ford body and enables hot

For the 75th anniversary of the 1932 Ford, a big celebration took place at the Grand National Roadster Show and also at Pebble Beach. All the well-known street rod builders were asked to build a pedal car that would be auctioned off at the show. Pictured are Chip Foose at left, Henry Ford III center, and Roy at right. The Brizio pedal car sold for $1,000 more than the Foose creation. *Roy Brizio Collection*

rodders to have a very nice coupe built without an original Ford body.

For more than 20 years, Roy Brizio Street Rods operated in the South San Francisco shop on Wattis Way. But space limitations caused the business to move across the freeway to a larger shop on Railroad Avenue in 2001. Today, Roy employs 15 people in the shop, and the crew produces one finished rod nearly every month. Roy Brizio Street Rods has built more than 300 cars, and Roy is quick to thank his crew of longtime employees for their hard work. With more than 30 years in the business, Roy Brizio is well known and respected in the street rod world. He makes dreams come true for his customers.

2 THE SHOP

This chopped three-window '32 is in the finishing stages and will have a nostalgic look, with an old-style engine, Lincoln drum brakes, white firewall and grille, and a tuck-and-roll interior.

THE BRIZIO SHOP in South San Francisco is one of the most active hot rod shops in the country, with 15 car-building experts building between 10 and 12 cars a year. Most of the boys in the shop have been with Roy Brizio Street Rods for many years, a few of them close to 30 years. That kind of experience makes for a good team, with many experts in the different disciplines of hot rod building. Most chassis buildup, body modifications, and all pre-assembly are done in the shop before the cars are sent to be painted. The cars return for final assembly and are then sent out for upholstery.

Under the lights in the shop, Roy's purple roadster looks blue. Roy has driven this roadster all over the place for over 14 years.

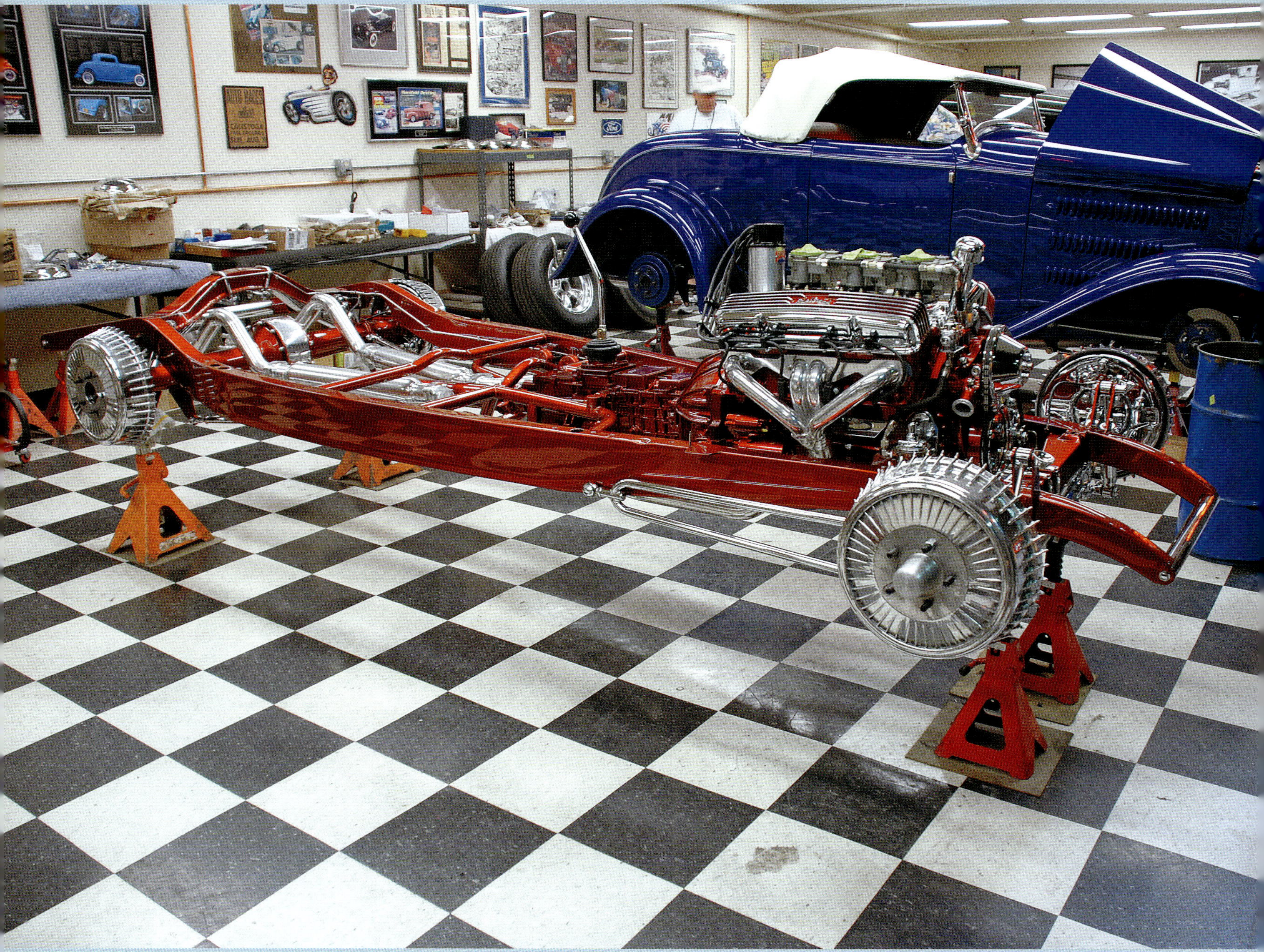

This copper-painted chassis with triple carburetors, a small-block Chevy engine, a five-speed Tremec transmission, and a Halibrand quick-change rear end is for J. J. Barnhardt's '32 five-window coupe.

Building a car is a long process, and thousands of man hours are involved. There are always between 25 and 30 street rods and 1950s cars in different stages of modification and buildup at the shop. It usually takes at least a year and a half to build a car, but with so many lined up in different stages, it's possible to finish a dozen cars each year.

With today's replica bodies in steel and some original 1930s Ford bodies still available, Brizio's is still able to build real steel cars. Some of the hard-to-find original bodies—such as the '32 roadster, '32 convertible, '32 three-window coupe, '34 roadster, and '34 three-window coupe—are today manufactured as identical steel replica bodies by different manufacturers. Many of the original bodies still used for building the cars are now more than 75 years old, which means they need some work, straightening them out or installing new wood. When it comes to the chassis, newer and safer frames are made, or the original frame is modified and the components are modernized versions of traditional pieces.

Brizio's team aims to give most of the cars a nice ride with a modern driveline, good working suspension with sway bars, coil-over shocks in the rear, big brakes, and yet an overall traditional old-Ford look. After chassis and body modifications are done and the

This custom '40 Merc was owned by John D'Agostino years ago; the new owner, Ted Stevens, bought the car at an auction and asked Roy Brizio to restore and update the chassis and driveline.

John Mumford's green 1934 Ford sedan delivery has a modern driveline. The suspension and brakes have been updated, yet John kept the old-style Ford look with light-green '35 Ford–style wire wheels.

car is pre-assembled, it is taken apart and shipped to one of the five Bay Area paint shops that Brizio's uses for finish work. With all parts painted, color-sanded, and polished (including the frame), a couple of Brizio's boys start the final assembly, piece by piece. With the car together again, the plumbing and wiring are done before the car goes to Sid Chavers in Santa Clara for upholstery. Before the customer can pick up his or her car, it is tested and fine-tuned during 250 to 300 miles on the road, and Roy Brizio ensures it is a good-running and safe car before it is delivered. Roy Brizio Street Rods is known for producing good drivers, but the cars are also very well detailed, and owners can be proud taking their cars to any car show.

Jim Vickery (left) and Tony Benedetti (right) are plumbing and putting together the chassis for a '32 roadster with a Chevy driveline and a chromed 9-inch Ford rear end. Coil-over shocks, ladder bars, and a rear sway bar will give the roadster a good ride. The roadster is owned by long-time customer Bubba Bugg.

Every little detail is laid out before final assembly. Each piece is checked when it comes back from the chrome shop, which is most often Sherm's Custom Plating in Sacramento.

A likely candidate for a magazine feature when it is done, this '33 roadster, owned by Larry Carter, is based on a Steve's Restoration body. It has some exotic parts, including a Ferrari engine with Weber carbs, a five-speed Tremec transmission, a quick-change rear end, and Halibrand-style mags.

Brizio's recently completed this '32 Ford Victoria with a blown Ardun Hemi engine for longtime customer Jorge Zaragoza of El Paso, Texas.

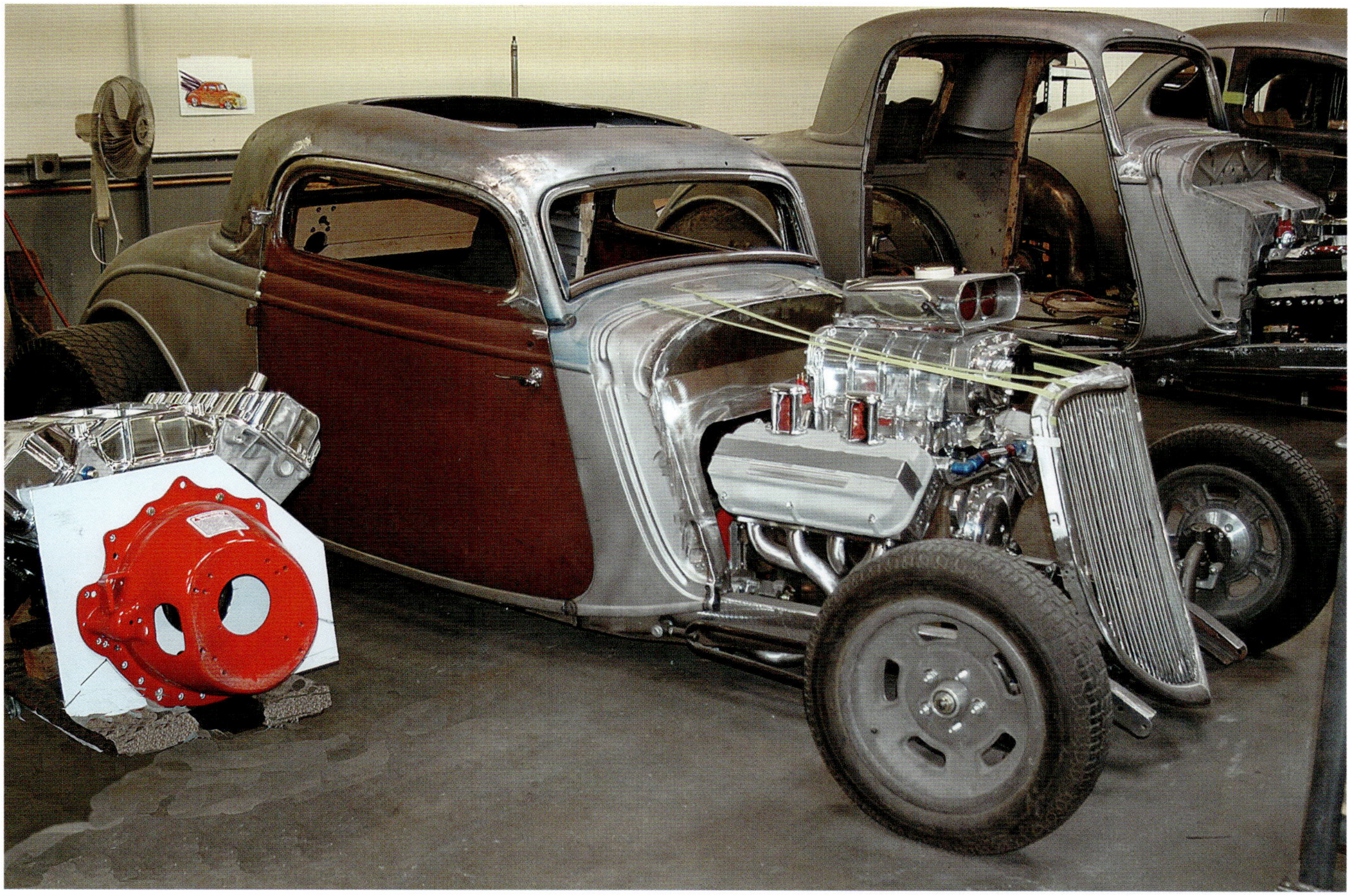

Brizio's built this wild '33 three-window coupe with a blown 392-cubic-inch Chrysler Hemi for Jorge Zaragoza. The body was slightly channeled, and the top was chopped with tilted A-pillars.

The '32 Vicky and '34 Tudor were media blasted and are now having new floors put in.

Roy Brizio has a knack for finding 1930s Fords, whether they are bodies or complete cars. This 76-year-old '32 Ford five-window body has been chopped a few inches and is ready to be dropped on a Brizio-made chassis. This car will have a blown Ardun flathead engine. It is owned by Mark and Dennis Mariani.

To find a '36 Ford roadster today is no easy task, especially on a budget. This body has been modified with a Bitchin firewall for more engine space. It is owned by Paul Bonderson.

Brizio's is working on more and more 1950s cars, and this '55 Chevy hardtop is just one of them.

Vic Edelbrock dropped off his '57 Chevy hardtop for some repairs.

Roy Brizio Street Rods was working on a '49 Ford coupe and a '49 Chevy pickup simultaneously for Eric Clapton. The truck has a brand-new replica body and a chassis with independent front suspension and a 9-inch rear end with coil-over shocks.

Brizio's engine room has all kinds of impressive hot rod engines. Note the fuel-injected small-block Ford in the foreground, a couple of 426 Hemis, and a ready-to-run flathead.

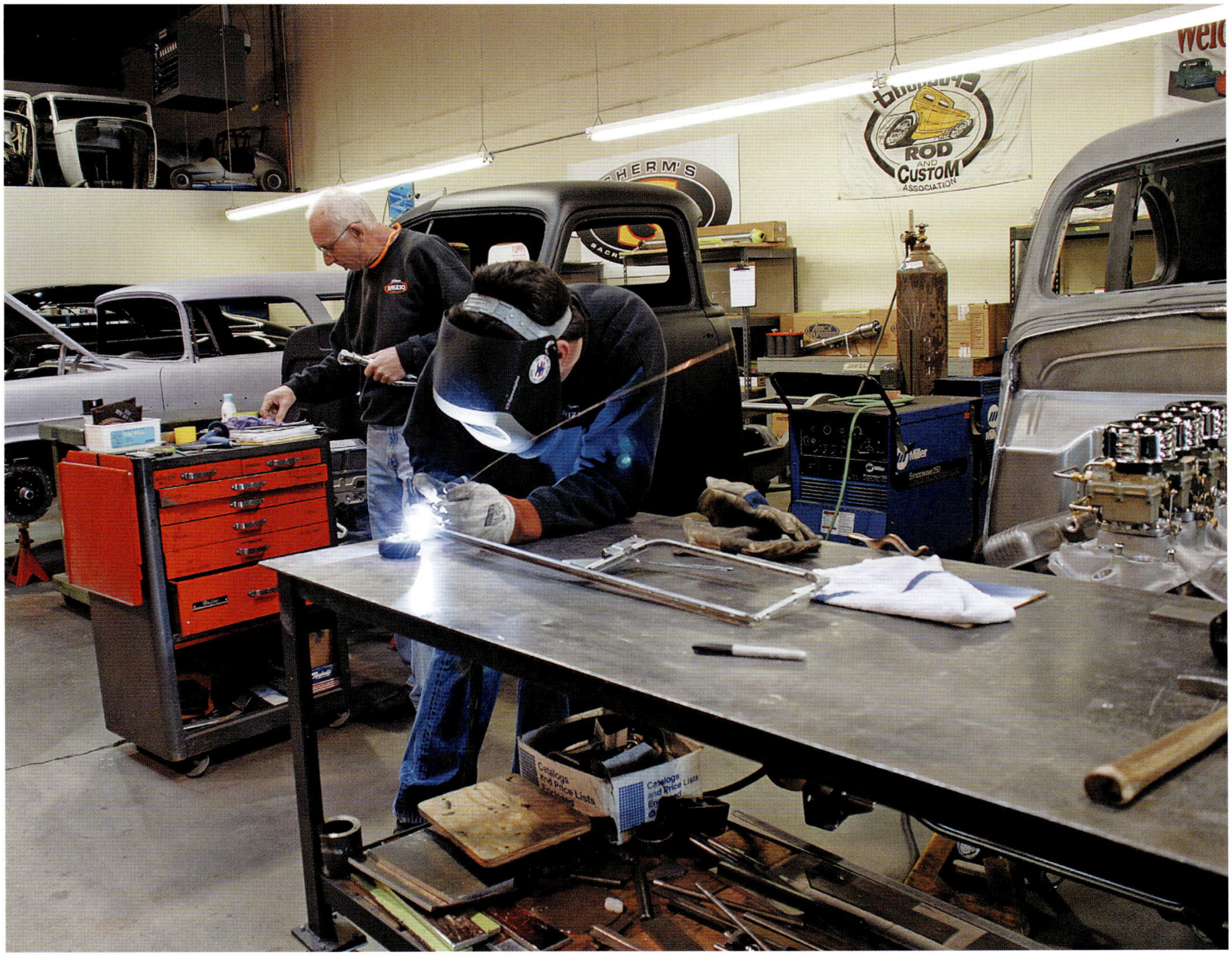

Roy gets a rebuilt, dyno'd, and ready-to-run Dodge Red Ram Hemi delivered for an upcoming project. It is a big job just to gather all the parts in time for each project.

Team Brizio

Standing (left to right): Joe Compani, Bob Lockwood, Steve Haas, Lenny Ernani, Jack Stratton, Dan Hall, Roy Brizio, Bill Ganahl, Jim Vickery, Paul Kriz, Gary Bammann, Steve Lord, Matt Connolly. Front row (left to right): Tony Benedetti, Paul Higgins, Dave Cattalini, Ryan Campi.

Steve Coonan

During the Open House, the backyard is full of hot rods—more than half of them are fenderless '32 Ford roadsters. Three rare full-fendered '32 phaetons are parked in the first row, while roadsters are traditionally parked in the back lot.

Open House

Once a year, on the Saturday before Mother's Day in May, Roy Brizio Street Rods hosts an Open House event attended by people from all over the West Coast. More than 200 street rods turn up with nearly twice as many people, and all guests are treated to free food, drink, and entertainment. Close to 75 cars will park in the lot behind the Brizio building; everybody else will park in front of the buildings along the street. Food is served in the backyard. People have a chance to take a closer look at all the ongoing and finished projects in the shop, and Roy Brizio has the chance to meet and talk with many of his customers, friends, and other visitors. The first Open House, held in 1986, was attended by some of the most famous people in the hot rod business, including Boyd Coddington, Fat Jack, Pete Chapouris, Jake Jacobs, and Pete Eastwood.

It is the same at today's Open House; you can always expect to run into some famous people from the street rod business, and the crowd more or less fills up the

After the Open House, the shop is full of extra Brizio-built cars that customers have left for the weekend.

shop. It takes plenty of preparation to host the Open House, especially to have lunch for 500 or 600 people, park all of the cars, and print extra T-shirts. Roy and his staff have plenty of practice from doing it every year for the last 22 years, and a lot of people look forward to the event each spring. Brizio's has more parking spaces now than at the old address, which makes it a little easier for everyone. Most people spend their time checking out the cars being built in the shop and the cars that the crew has already finished.

During the Open House, the shop is open for visitors to see all the projects in process.

All three '32 five-window coupes are unchopped, and the nearest one is all stock. Both it and the coupe with the candy blue paint and white interior are Roy's cars.

The front of the Brizio building is packed with cars, including this custom, sectioned '41 convertible which has been owned by Harry Costa for more than 50 years.

Brizio's oldest customer is Andy Sargenti, 92. Andy bought this '54 Ford semi-custom convertible new in 1954. He had the Ford customized in the 1960s with Frenched headlights, shaved handles, and rounded edges on hood and trunk lid.

3 MODEL A's AND T's

Like most of Roy Brizio's work, the Carlos Lara Model A roadster is not a show piece—it is very much a driver and has a lot of unusual parts, including a Jag motor with Weber carburetors. *Steve Coonan*

The *Ala Kart* was a big success as soon as it was shown at the 1958 Grand National Roadster Show. Richard Peters toured with it, and the model company AMT showed it for some years in the 1960s. More than one million model copies were sold.

EVEN THOUGH ROY BRIZIO Street Rods is best-known for its '32 and '34 Fords, the shop has also built plenty of other beautiful machines over the years, including models that predated the Deuce. One of Brizio's finest achievements was a restoration of the famous *Ala Kart* that George Barris built for Richard Peters in 1958.

Hershel "Junior" Conway did the finish work on the original car and painted it in white pearl. George Barris topped Junior's paint job with purple and gold scallops. *Ala Kart* won the prestigious America's Most Beautiful Roadster award at the Grand National Roadster Show in 1958 and 1959, prompting toy model kit company AMT to lease the car from Peters and eventually buy it to use at car shows. However,

Richard Peters took the Dodge Red Ram engine out of his race boat for the *Ala Kart*. The car was featured on the cover of *Hot Rod* magazine in 1958 with a Hilborn fuel injection system. The restored 266-cubic-inch Hemi motor was finished the same way.

It was a great moment at the Grand National Roadster Show when the fully restored *Ala Kart* was uncovered in 2008. It was accompanied by a complete replica of the display from the 1958 show, including the car's setup with foil-covered stands and a mirror under the car.

The original *Ala Kart* was shown at the 1958 show on a white carpet, with a big mirror under the chassis and foil-covered stands. The car's original owner, Richard Peters, was at the 2008 show to see the restored car.

in 1963, the car had an electric short that resulted in a fuel line fire. The damaged icon seemed to disappear for a long time.

Then, in 2001, one of Junior Conway's customers turned up at his shop with the old *Ala Kart* and asked him to restore it. Junior started by stripping the sheet metal down to bare metal and finishing it with new lead. After Conway had finished the pickup bed and some of the body pieces, the owner decided to sell the car. Roy Brizio and Steve Coonan convinced John Mumford to buy it. The car was eventually delivered to Brizio's shop, and Roy put one of his guys, Bill Ganahl, in charge of the project. Bill invested countless hours in research and physical work to prepare the *Ala Kart* for the Grand National Roadster Show and consideration for the America's Most Beautiful Roadster award.

The *Ala Kart* was based on a stock '29 roadster pickup. Peters modified the frame with chrome brackets for coil springs in all four corners. He had the chassis, body, pickup bed, and fenders delivered to the Barris shop in Lynnwood. Peters had removed a '54 Dodge Red Ram Hemi from his race boat and bolted it to a '39 Ford transmission with Lincoln gears. The rear end came from a '32 Ford with bolted-on '40 Ford hydraulic brakes. Peters installed a dropped I-beam axle on the front with '39 Ford spindles and '40 Ford brakes. At the Barris shop, the crew modified the body with a '27 Model T back panel, shortened and modified the pickup bed, bobbed the fenders, and made a new front end with a round grille and dual headlights.

It was a big moment when Roy Brizio took the *Ala Kart* out for a spin; few people have seen this famous old show car on the road.

A new aluminum hood was fashioned with a sunken hood scoop. Conway, who was working for Barris at the time, finished the paint job, and Roy Gilbert crafted the upholstery and special seats.

For the restoration, Roy's crew repeated the original buildup bolt for bolt. Mickey Galloway did the metalwork, Darryl Hollenbeck and Art Himsl painted the car, and Howdy Ledbetter recreated the interior perfectly. The restored *Ala Kart* debuted at the 2008 Grand National Roadster Show in a display just like that used in 1958. George Barris and the car's original owner, Richard Peters, attended the show to commemorate the car's return.

The swing pedals came from a '54 Ford, and the dash panel has a set of '57 Corvette gauges, just like when it was first built. The steering wheel, with a gold bullet in the center, came out of a '57 Lincoln.

Darryl Hollenbeck copied the original Junior Conway Barris paint job down to the smallest detail, and Art Himsl matched the purple and gold candy colors. The same scallops are painted on the underside of the fenders.

After a wiring and fuel fire, the *Ala Kart* was parked in a garage until 2001, when Junior Conway began its restoration. *Steve Coonan*

A '32 Ford rear end was chromed, along with many of the other chassis parts. The coil springs and air bags were new ideas in 1958, and the four pipes sticking out under the Frenched license plate added to the custom style.

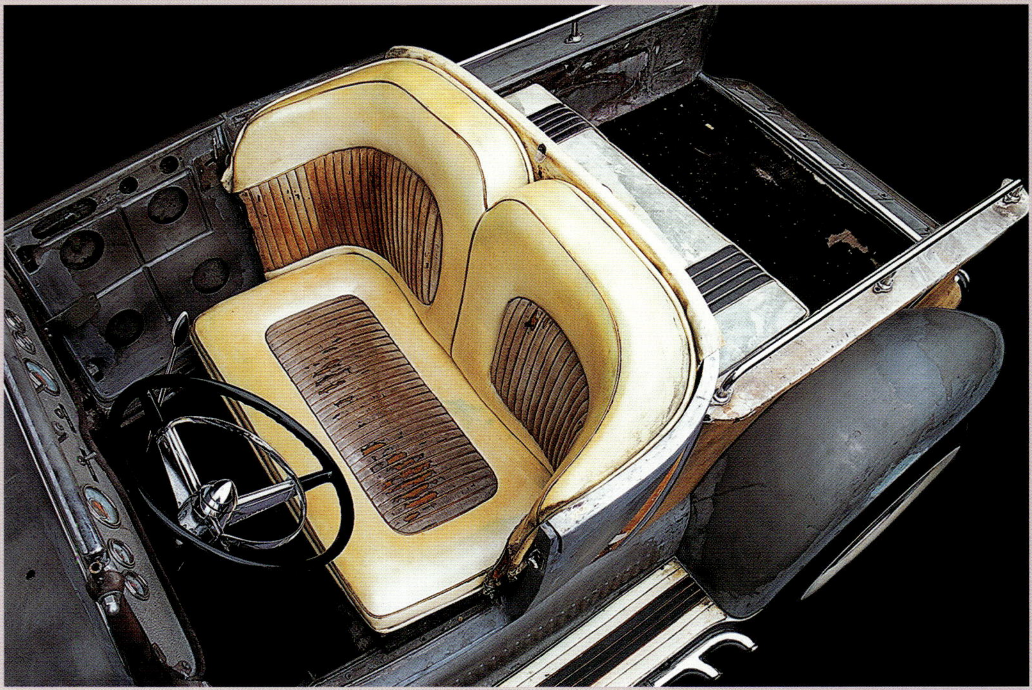

In this prerestoration picture of the *Ala Kart*, some of the upholstery is still intact, which made it easier for Howdy Ledbetter to replicate the original upholstery.
Steve Coonan

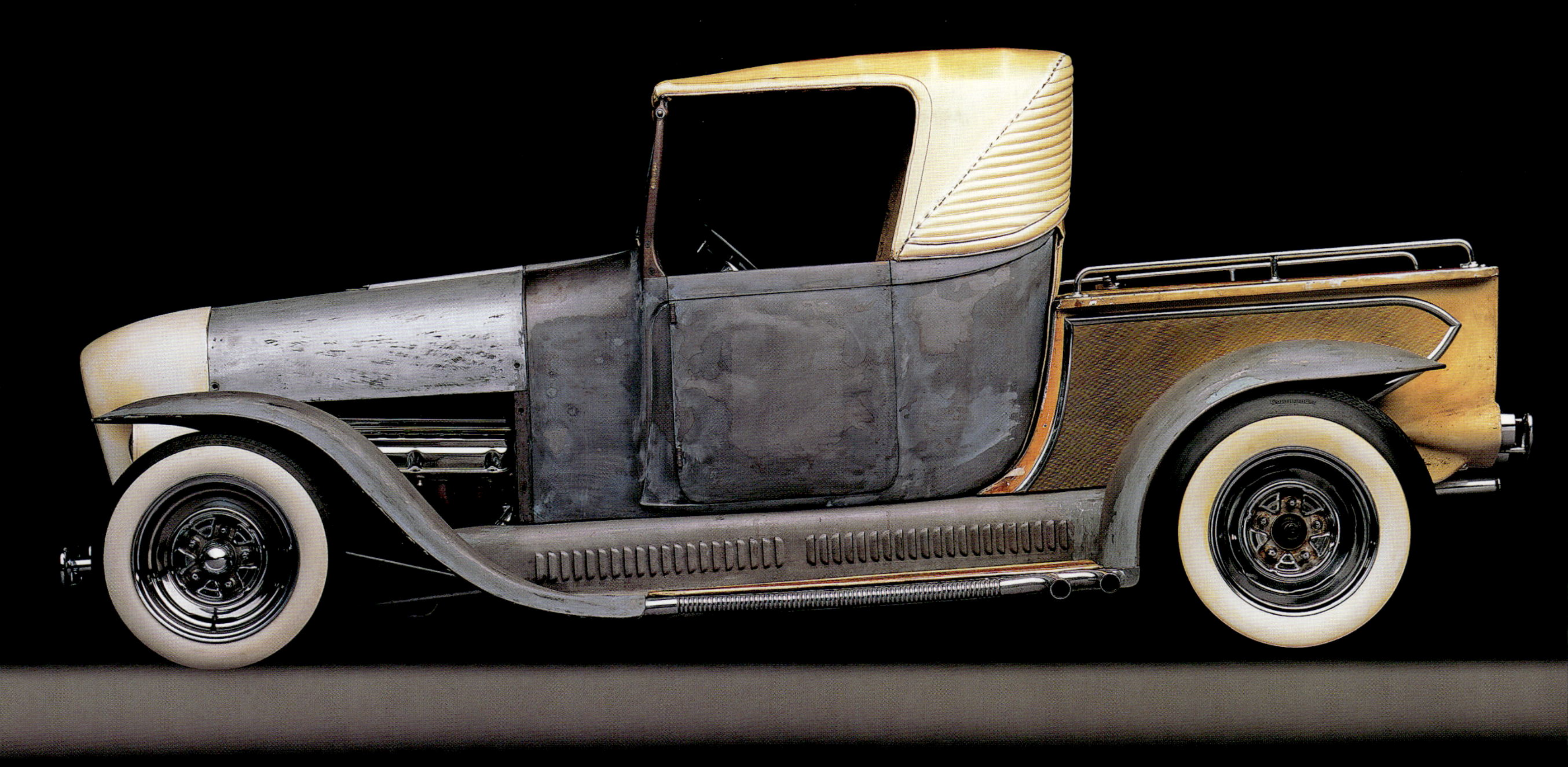

Junior Conway had started to restore the *Ala Kart* for one of his customers before it was sold to John Mumford. This photo shows how the car looked when the Brizio team took over and completed the restoration. *Steve Coonan*

The *Ala Kart* was nearly all intact, parked in a garage for more than 30 years before it was sold and Junior Conway was hired to restore it. This street rod gave George Barris notable publicity; it won the America's Most Beautiful Roadster award in 1958 and 1959. *Steve Coonan*

Following spread: Two very famous Sam and George Barris–built cars are back together after all these years. The *Ala Kart* Model A roadster pickup appears with the Sam Barris '49 Mercury, which also can be seen in Chapter 6. *Steve Coonan*

With the new hood a little longer than stock and the 1932 frame pinched in to be cut and hidden behind the grille shell, the front end gets a very clean look. The grille and grille shell are new replicas in steel.

Cliff Hanson's '31 Roadster Pickup

Cliff Hanson of Zephyr Cove, Nevada, wanted a fenderless roadster pickup and sat down with Roy Brizio to get his ideas on paper. They put together plans for a '31 Model A Ford roadster pickup.

Brookville Roadsters makes an open steel-body replica of the Model A, so Cliff's project began with an order for a pickup body and shortened bed from Brookville. Roy and Jack Stratton used a set of American Stamping '32 frame rails to build the chassis. New tubular crossmembers were made for the transmission bracket, and the front of the frame was pinched to fit behind the '32 grille shell. The front crossmember was also flattened to get maximum front-end drop.

The front end has a 5-inch dropped Super Bell axle and a mono-leaf spring held in place with a set of Pete & Jake's hairpins. The Dutchman quick-change rear end has an early Ford axle with hairpins and a set of coil-over shocks. The wheelbase was stretched to 110 inches—4 inches longer than a stock '32 Ford.

With the basic chassis complete, a Mullins Vega steering box was paired with Pete & Jake's shocks and Panhard bars. Late-model Ford rear drum brakes work in combination with a set of So-Cal Speed Shop disc brakes hidden in finned, Buick-style drums. The roadster is powered by a 350 Edelbrock crate small-block connected to a five-speed Tremec transmission with overdrive and a 4.11:1 gear ratio. The Edelbrock engine was delivered with 10.5:1 compression; Edelbrock aluminum heads, cam, and intake manifold; and dual Edelbrock four-barrel carburetors under a Moon air cleaner. With

The stance of the car, with its front-end drop, screams "hot rod." With the blue candy paint, red and white accents, and the wide whitewalls, it's easy to see why people love this car.

More and more hot rod builders do what Cliff did for his car: order a brand-new crate motor to get some fresh horsepower. Cliff selected an Edelbrock-built, 410-horsepower 350 Chevy with plenty of goodies, including aluminum Edelbrock heads and cam, a polished intake manifold with dual four-barrel 500-cfm carburetors, and a Moon air cleaner.

Roy Brizio is one of the best in the business at building good-looking drivers, and most of them have a Sid Chavers stitched interior like this one in red-and-white, 1950s-style tuck and roll. The 1932-style dash has a Brizio insert and a set of Classic Instrument gauges. Because a hot rod should be driven like a sports car, owner Cliff Hanson wanted a five-speed manual with a Hurst shifter.

410 horsepower, Cliff has all the hot rod power he could want.

Even a brand-new steel body needs finish work before it can be painted. Mickey Galloway took care of the body and pickup bed, and the aluminum hood was then crafted. In Concord, Darryl Hollenbeck painted the bodywork in House of Kolor blue candy. With the car bolted together, the chassis parts were chromed and painted, and the wiring was installed. The car was then trailered to Sid Chavers—Brizio's main man for interiors—in Santa Clara. Sid stitched a 1950s-style interior in red-and-white Naugahyde, and local pinstriper Rory Pentecost pinstriped the exterior. Cliff showed off his hot little pickup at the L.A. Roadster Show, and among the 900 roadsters at the show, Cliff's was a crowd favorite.

The deeply dropped front end has a 5-inch dropped Super Bell tube axle with Pete & Jake's spindles, hairpins, and shocks. The Buick-style So-Cal Speed Shop brakes have hidden disc brakes inside.

The car was based on the old lakester style with some modern touches, including a Jaguar engine and a set of big American Racing 17x20-inch wheels with low-profile BF Goodrich tires.

Carlos Lara's '29 Lakester

Hot rodder Carlos Lara moved from Mexico to Alberta, Canada, just before he contacted Roy Brizio Street Rods about building a roadster with a set of Deuce frame rails and a replica Brookville 1929 roadster pickup body. Carlos delivered a Jaguar six-cylinder engine to power the hot rod.

Classic lakesters were built on Model A or '32 chassis with Model T bodies or just a basic '29 roadster pickup body (without the pickup bed) with a tank in back on top of the frame. Roy Brizio and his crew formed the '32 frame rails after the body, and the rails were cut in the front and rear for a bobbed style. While the shop was working on the chassis, Jag specialist Jim Griffin took apart the 1968 4.2-liter XKE engine and put it back together with a high-lift cam, a ported and polished head, and Isky valve springs. Three chromed 40-millimeter DCOE side-draft Weber carburetors from Inglese were mounted on a Jaguars Unlimited intake manifold. A Mallory Unilite distributor was selected for ignition, and a Tremec five-speed transmission was installed with a Brizio-made bell housing. Power is fed to a Dutchman quick-change rear end with 3.50:1 gears and a high Model A–type leaf spring from Betts Spring Company in Oakland. The front end was built

To keep a clean look, everything possible on the 1968 Jaguar XKE engine was polished or painted silver. Jim Griffin completely rebuilt the 4.2-liter engine. With 9:1 compression, high-lift cams, and a ported head, the engine was topped off with three Weber carburetors on Jaguars Unlimited manifolds. The straight-pipe exhaust was installed for show, and it looks very good with the rest of the polished valve covers and carburetors.

The lakester has a quickchange rear end built by Dutchman, with 3.50:1 gears. The aluminum pieces were polished before the rear end was assembled, and it is held in place by a set of Pete & Jake's ladder bars, plus a Model A leaf spring.

The Sid Chavers interior was stitched in tan Jaguar leather, and the seats were kept slim to give owner Carlos Lara some extra leg room in the car. The 1932-style dash is filled with Classic Instrument Hot Rod Series gauges, and the Hurst shifter is hooked up to a Tremec five-speed transmission.

The old race car–style look included a set of Speedway Motors friction shocks in the front and 1929 Model A headlights with modern halogen inserts. The headlights were mounted extra low and out in front of the Model A grille shell.

on a 5-inch dropped Super Bell tube axle with Pete & Jake's radius rods and a monoleaf front spring. For an old-style racing look, Brizio selected a set of Wilson Welding finned backing plates and Buick aluminum drums with Ford hydraulic brakes all around.

To give Carlos some extra leg room in the little '29 Ford roadster body, the Brizio boys dropped the floor under the seats and used a slim backrest. Kustom Classics painted the body in DuPont British Racing Green, and Jim Vickery wired the car. The '32-style dash has Classic Instrument Hot Rod Series gauges, and the three-spoke Bell-style steering wheel was wrapped in tan Jaguar leather by Sid Chavers. Sid stitched the entire interior, including the carpet, with matching leather edges. The lakester rides on American Racing 17- and 20-inch wheels with low-profile 205 and 295 BF Goodrich tires. The Model A has six short straight pipes, making for a very short exhaust system—great for shows, but not so great for the street.

Jim Stroupe owns this little '27 T original-bodied roadster with a narrowed body, 1960s Alfa Romeo engine, four-speed transmission, Halibrand quick-change rear end, and rare Kinmont brakes all around. The owner had the parts and the ideas, and he then hired Brizio's to build his dream car.

Brizio's built the little T with a square tubing chassis and a Wescott '27 T fiberglass body. Under the body were a Halibrand quick-change rear end, a 302-cubic-inch Ford engine with dual Edelbrock four-barrels, and a T five-speed transmission. The tube front axle had coil-over shocks fitted inside the track-nose front end.

Chip Foose came up with the idea to build a clone of the first *Hot Rod* magazine cover car to celebrate 50 years of the magazine. Foose also made the rendering of the updated version of Regg Schlemmer's '27 T roadster, which Brizio built in 1998.

4 '32 FORDS

This picture of Glenn Grozich's Brizio-built '32 coupe was taken with the San Francisco Bay in the background. Roy Brizio knows how to mix old traditional parts with today's high-performance parts, including a 600-horsepower 514-cubic-inch Ford big-block, a six-speed Richmond transmission, and 17- and 20-inch Billet Specialties wheels.

There are not many unchopped three-window street rod coupes around these days, but Jeff Beck's is pretty near stock, except for the top that was filled under the white cover. The combination of '57 Chevy Sierra Gold paint, whitewalls, the white top cover, and the running boards screams early 1960s.

Boogie-Woogie Three-Window

I first met Jeff Beck in 1975. I had heard that he had some hot rods in his garage in Sussex, England, and he was nice enough to show me his collection. He was doing a lot of the work himself, and he had just finished an Andy Brizio–style T-bucket with a blown small-block Chevy.

"Jeff has become a good friend through the years, and we have delivered a few chassis and other parts for his cars," says Roy Brizio. This latest project started when Roy and his wife visited Jeff in England a few summers ago. Jeff told Roy, "I should send this old three-window over to you, and let you build me a neat street rod." About a year later, Jeff shipped

A brand-new crate 350 Chevy was painted and detailed with Mooneyes valve covers and air cleaner for Jeff Beck's coupe. Between the heads, a polished Edelbrock manifold is paired with an Edelbrock carburetor.

The new chassis was painted white, and it has a dropped I-beam front end with a single leaf spring, a pair of hairpins, and a set of chromed Aldan shocks. The '39 Ford-style spindles have a set of Super Bell disc brakes behind the chromed steel wheels with '50 Merc caps.

The 15-inch 165x15 and 235x15 radial whitewall tires are mounted on chromed steel wheels with '50 Mercury hubcaps. The front fenders are new '32 Ford steel reproductions, and the rears are fiberglass replicas.

the body to San Francisco, and Brizio's began to plan the project. Jeff wanted to build a late-1950s-, early-1960s-style street rod, similar to Roy's purple '32 roadster.

The Brizio crew built a new chassis with a dropped I-beam front axle, hairpins, and a dropped spring. The 8-inch Ford rear end is held in place by long ladder bars, Aldan coil-over shocks, and a Panhard bar. A brand-new 350 Chevy crate engine with an Edelbrock manifold and a Mooneyes air cleaner and valve covers powers the street rod. Brizio's bolted on a 700R transmission, and Jack Stratton fashioned a stainless-steel exhaust system. The body needed little work, but it was straightened before it was dropped on the chassis. The front fenders and tank are new steel reproductions, while the rear fenders are fiberglass. Dale and Darryl Hollenbeck did the finish work and painted the rock-and-roll coupe in '57 Chevy Sierra Gold.

Brizio's assembled the completed frame, body, and other parts, and Jim Vickery did the wiring. Sid Chavers crafted a classy pearl tuck-and-roll interior. I witnessed the finishing touch, when Rory Pentecost added Tommy "the Greek" Hornes–style pinstriping.

Sid Chavers, who does most of Brizio's interiors, stitched this one, too. The pearl white tuck and roll gives the car plenty of light on the inside.

The high-boy Tudor was well built by Brizio's and boasts an 850-horsepower blown big-block under the hood. Frank also had a chromed roll cage installed in the car. The Tudor looks wild coming down the street, with the blower scoop sticking out of the hood and the headers along the hood sides.

Frank Ghiglione's Blown Tudor

In 1962, Bob Tindle built the *Orange Crate*, a fenderless race-type '32 sedan with an Olds engine and a front-mounted 6-71 blower. That car served as inspiration for Frank Ghiglione's fenderless high-boy Tudor sedan. Frank's Alameda, California, garage houses an impressive mix of more than 50 cars, from street rods and customs to 1950s classics and muscle cars.

In 1980, Frank bought a '32 Tudor sedan for $4,500 but stored it until he could work on it. Some years later, he hired Roy Brizio to do the job. Roy and his team put the '32 frame rails in a jig, with new crossmembers in place, and mounted the front and rear ends. The front end is a traditional 5-inch dropped Super Bell axle with hairpins, new replica '39 Ford–type spindles, steering arms, short chromed shocks, and a reversed spring. Frank chose a Halibrand quick-change rear end with 3.55:1 gears. The rear end was mounted with a set of long Brizio-made ladder bars and a pair of coil-over shocks.

For the engine, a brand-new 502-cubic-inch marine Chevy big-block was bored to 510 cubic inches and put together with a set of 8.75:1 blower pistons, a

Howdy Ledbetter of Fremont crafted the black leather interior and upholstery over a pair of Teas seats with a little Howdy-made back seat. The steering column is attached to a Mullins Vega steering box and a Pete & Jake's Bell-type steering wheel. The gauges are modern Mooneyes in the old style.

The Halibrand quick-change rear end was mounted with a set of Brizio-made long ladder bars and coil-over shocks. The gears are 3.55:1, but Frank can change them quickly if he wants to put slicks on and go for a run at the drag strip. A set of disc brakes in the rear gives the car plenty of braking power.

The engine and transmission were built by B&M with a brand-new 502-cubic-inch engine as a base. It was bored out to 510 cubic inches and put back together with 8.75:1 blower pistons and a Crane cam. The heads are iron Merlins with roller rocker arms, and a B&M Mega blower with two 750-cfm Holley carburetors tops the manifold under the scoop. The engine produces 850 horsepower at 5,000 rpm.

Frank Ghiglione is a trucking company owner and car collector from Alameda, California, who loves wild street rods. The Roy Brizio–built high-boy Tudor sedan is the best known of Frank's collection.

Crane cam, and a set of Merlin iron heads. Between the heads, Brizio's team installed a B&M manifold and a B&M Mega blower with two 750-cfm Holley carburetors under the scoop. A TH 400 transmission was built to harness the 850 horsepower of the engine.

Marcel's Custom Metal in Corona reworked the body after the chassis took shape. Sheetmetal magician Marcel DeLay chopped the top 2 1/2 inches, filled the roof, and widened the wheelwells. After the body was final fit to the chassis, Toy Works in Concord did the finish work and painted the parts in DuPont orange. In typical northern California style, Rory added black and silver pinstriping along the body line and on the louvers. Jim Vickery wired the car before it was sent to Howdy Ledbetter to cover the interior in black leather, including the door panels and headliner. Completed in about 12 months, the high-boy Tudor rolls on a set of Halibrand Swirl wheels with Michelin and M/T tires.

Roy Brizio and his crew are used to building easy-to-drive street rods for their customers. Longtime employee Matt Connolly drives this great-looking coupe for the rolling shots.

Billet Specialties Five-Window

Glenn Grozich of Burr Ridge, Illinois, runs Billet Specialties, a company that makes machined aluminum parts. When Glenn was 14, he became obsessed with owning a Deuce coupe. One day, the teenager saw an ad in the paper: "1932 5-window coupe for sale, 1,500 dollars." The seller received so many calls that he told Glenn he would open his garage door at 7 a.m. the next day and sell it to the first guy with the money in his hand. Glenn was there at 5 a.m. He brought home a body and frame and spent three years finishing the coupe with the help of his friends. The payoff came when he showed up at his high school's graduation party in his yellow *American Graffiti*–style coupe with a 302-cubic-inch Chevy. In 1982, he parked the coupe and turned his attention to a '69 Camaro.

Years later, during a conversation at the Nationals with Roy Brizio and Angelo Giampetroni of Ford Racing Performance Parts, the idea of bringing out the coupe and rebuilding it came up. Soon afterward, Glenn shipped the coupe to Roy Brizio Street Rods in San Francisco. Aside from the body, little of the old car would remain. Brizio's built a new frame in the jig, extended the wheelbase 2 inches to 108 inches, and gave the rear of the frame a higher kick-up. The front end is a 5-inch dropped Super Bell tube axle with hairpins and a Durant single leaf spring. The rear end is a shortened Ford 9-inch with a set of long Pete & Jake's ladder bars and a pair of Aldan coil-over shocks. For better handling, a Panhard bar was installed up front with a 3/4-inch Deuce Factory sway bar in the rear. Large Wilwood disc brakes counter the power under the hood.

The only part left from Glenn's old coupe is the body, which he bought when he was 14 years old. Jim Hendricks at Brizio's did the bodywork, including chopping the top 2 inches, filling the roof, making a rolled rear panel, and straightening out the old steel before the paint job.

Brizio's detailed the Ford Racing Performance Parts engine with paint and polished the manifold. The top end was completed with a 1,050-cfm Holley carburetor and a Billet Specialties air cleaner in aluminum. To keep the hot big-block cool, a Griffin radiator was paired with a Vintage Air electric fan.

As for that power, Glenn, Roy, and Angelo decided on a new Ford 460-cubic-inch big-block, stroked to 514 cubic inches, delivering 600 horsepower out of the box. Brizio's combined the engine with a Weber clutch, Lakewood bell housing, and a brand-new six-speed Richmond transmission. Dan Hall and Jack Stratton made headers and a 2 1/2-inch stainless-steel exhaust system with a pair of Stainless Specialties mufflers. Jim Hendricks chopped the top 2 inches, filled the roof, and made a rolled rear panel. Jack Hagemann fashioned the hood and put plenty of louvers in the hood sides for ventilation.

The finish work was done by Paul Camilleri's Auto Works in San Carlos, not far from Brizio's. Everything was painted in the House of Kolor Lime Time green candy, and Art Himsl painted the flames and

The dash was updated with a Billet Specialties insert filled with Auto Meter gauges. The steering column and the steering wheel were also made by Glenn Grozich's company, Billet Specialties, and the column hooks up to a Vega steering box. The Hurst shifter is bolted onto a six-speed Richmond transmission.

pinstriping. When all of the mechanical parts were in place and Jim Vickery had completed the wiring, the coupe was transported to Sid Chavers in Santa Clara for black leather upholstery with a design in the panels. Glenn dubbed the car *Pro-Rod*, and I had a chance to go for a ride in the attention-getting coupe.

The car was restored down to the smallest detail, including the drilled split wishbone, the drilled I-beam axle, the L.A. Roadster plaque in the rear, and the American Racing rims and Firestone bi-ply tires.

Tom McMullen's Legendary Roadster

Many replicas of this car have been built, and Tom McMullen himself has even built a few different versions of it through the years. Tom founded *Street Rodder* magazine, which still uses the roadster as its symbol. He bought the car in 1958 with a 283 Chevy small-block, a '39 Ford transmission, and a '39 Ford rear end. Tom and the roadster became regulars at local drag strips, and he built a 352-cubic-inch stroker small-block with six 97 Stromberg carburetors for plenty of power. He drove the car on the street and the drag strip, burning through a few engines.

The car had several paint jobs, including one in then-experimental green metalflake. Eventually, Tom painted the car black himself and took it to Ed "Big Daddy" Roth for some special treatment. Roth taped the flames, Tom painted them, and then Ed pinstriped them and added white pinstriping all over the car. By that time, the car was powered by a 301-cubic-inch small-block with a 4-71 blower and two four-barrel carburetors. When the roadster appeared in *Hot Rod* magazine in 1963, it had a Moon tank in the front and a full set of Stewart Warner gauges in the dash, and it rode on American Racing wheels. The car became famous because Tom drove it daily and raced it at the drag strip and the lakes. He continually modified the roadster, boxing the frame, installing bigger engines, trying out four-speed and automatic transmissions, and building a quick-change rear end.

When present-owner Jorge Zaragoza bought the old roadster, he asked Roy Brizio to restore it to the 1963 style, including the 301-cubic-inch Chevy, '39 Ford transmission, and closed drive with the '39 Ford/quick-change rear end. The team did a lot of research to determine exactly how the car was built in those days. The legendary roadster now appears just as it did in 1963, with a black-and-white Naugahyde interior, flames, and pinstriping.

Tom McMullen's roadster is one of the most well-known hot rods in the country. Brizio's brought back the original look of the car, and the flames were copied from the original design that Ed Roth taped and Tom painted. Roth outlined the original flames with pinstriping. For the restored car, Darryl Hollenbeck replicated the paint job and Art Himsl painted the flames.

The interior is done in black-and-white tuck and roll. The dash was filled with a full set of old-style Stewart Warner gauges, and the steering wheel is an old Bell type. The tonneau cover was made for racing and to keep the wind passing smoothly over the seat area.

The 283-cubic-inch Chevy small-block was bored and stroked to 301 cubic inches with blower pistons, a special cam, and a 4-71 GMC blower between the heads, plus two four-barrel Rochester carburetors on top—just the way Tom McMullen had it in 1963.

Even Ed Roth's early 1960s pinstriping was re-created by Rory Pontecost. The car also has the original license plate number.

The rake of the rod is just as perfect as can be, thanks to the dropped front crossmember and I-beam axle. The pickup bed was ordered from Brookville Roadsters 18 inches shorter than stock, and fit the car just right. The Brizio team fitted the sheet metal and made the hood and some of the pieces to mount the pickup bed.

Poteet's Roadster Pickup

George Poteet of Memphis has a bunch of nice cars in his garage, including one other hot rod built by Roy Brizio Street Rods. For his latest project, George sat down with Roy to discuss a hot rod pickup. George suggested a '32 Ford roadster pickup based on a Brookville Roadsters steel replica body and an 18-inch shortened pickup bed. The brand-new replica steel body was easy to work with and saved lots of time. The frame is built in straightforward hot rod style, with new American Stamping frame rails, a lower Model A front crossmember, a tube center crossmember, and a stock-type '32 rear crossmember. A dropped I-beam axle is held in place by a pair of hairpins, and the front suspension is made up of '39 Ford–style spindles, a single leaf spring, and a pair of chromed Pete & Jake's shocks. A Mullins Vega

Both front and rear suspensions are held in place by old-style Brizio-made hairpins, and special covers were made for the tie rod ends.

With 515 horsepower under the hood, George Poteet needed plenty of brakes. A set of Wilwood performance disc brakes with drilled rotors both front and rear provide all the needed stopping power.

steering box was paired with a 9-inch Ford rear end cut to size by Currie Enterprises. The rear suspension is a '40 leaf spring with a pair of hairpins and chromed shocks. To adequately stop his powerful roadster, George chose a set of Wilwood performance brakes with drilled rotors for all four wheels.

George wanted a hot engine under the hood and selected a Roush-built 402-cubic-inch Ford crate engine. It was based on a 351-cubic-inch engine that was bored and stroked to 402 cubic inches with special ported Roush aluminum heads and intake manifold, and a 780-cfm Holley under a Billet Specialties air cleaner. The hot engine was dropped in the chassis with a new Tremec five-speed transmission shifted with a Hurst shifter.

George wanted his roadster to ride on old-style "kidney-bean" wheels, which are now reproduced. The magnesium Speedway wheels are 6x16 inches in front and 7x18 in the rear, and are shoed with old Dunlop racing tires. Paul Camilleri performed the finish work and painted the car in DuPont black before it was pinstriped by Rory Pentecost. As with most Brizio projects, Sid Chavers stitched the leather interior and made the convertible top. The roadster pickup rolled out of the San Francisco shop just in time for the L.A. Roadster show at Pomona.

Sid Chavers, who does most of Brizio's projects, stitched up the interior. The upholstery is beige leather, and the dash has an Auburn-style Kinectic dash insert with Classic Instrument gauges. The Mullins steering column is topped with a Budnik sprint-style steering wheel.

The engine and transmission combination is really hot in this hot rod because Poteet ordered a Roush-built 402-cubic-inch Ford motor that pumped out 515 horsepower on the dyno. An MSD distributor takes care of the ignition, and the Holley 750 carburetor pumps in the air/fuel mix. Transmission is a new five-speed Tremec.

Roy changes wheels on the roadster now and then, and he swapped the '50 Merc wheels with the whitewall tires for a set of five-spokes with blackwall tires. The roadster is more than 14 years old and still looks like new, even with plenty of miles on the odometer. The 302-cubic-inch high-performance small-block Ford engine makes this hot rod pretty quick.

Toy Works painted this good-looking roadster in Ford Mustang purple. This roadster was one of the first built in 1960s nostalgia style, with a white interior, running boards, and whitewall tires.

Roy Brizio's Purple Roadster

Built in the mid-1990s, Roy Brizio's purple '32 roadster has seen more road miles than most hot rods. The project began after Roy talked with Angelo Giampetroni from Ford Motorsports about building an all-Ford based '32 roadster. Roy built the chassis and topped it with a Wescott fiberglass body. To pull it all together, the car was given an old-fashioned look, with a bright candy paint job, a white interior, whitewall tires on chrome reversed '50 Mercury wheels, and a peppy Ford small-block with a five-speed manual transmission.

Jack Stratton set up a pair of American Stamping '32 Ford frame rails in the frame jig and added a Model A–style front crossmember, a tubular center cross-member, and a straight rear crossmember. A dropped Super Bell I-beam front axle, four-link, lowered front spring, Panhard bar, and pair of Pete & Jake's shocks composed the front suspension. A Vega-style steering box and cross tie rod were connected to a Mullins steering column with a Bell-style replica steering wheel. The Ford 8-inch rear end suspension is set up with long ladder bars, Aldan coil-over shocks, a Panhard bar, and a sway bar to keep the rear tires grounded. A brand-new high-performance 302-cubic-inch Ford GT 40 small-block with two Holley four-barrel carburetors gives Roy all the power he wants.

With the engine and a Tremec five-speed transmission in place in the chassis, Brizio's team made a stainless-steel exhaust system and installed the plumbing before the full-fendered fiberglass body was lifted onto the chassis. Like most Brizio-built '32s, Roy's roadster features a grille shell cut down 1 inch and a new hood to fit the drop-down angle. In keeping with the car's old-style look, nerf bars were made for the front and rear frame horns.

Brizio's completed the dash with a set of Ford Classic gauges in a Lobeck-made Stewart Warner–style insert. As with all Brizio-built hot rods, the car was put together and then disassembled for the finish work. When the parts returned from paint and chrome work, the team already knew everything would fit together. Toy Works performed the prep work, including several hours of sanding and filling, and they painted the hot rod in a special Ford Mustang blue-purple DuPont candy. With the car bolted together again, Jim Vickery did the wiring and Sid Chavers stitched a pearl white interior and running board covers.

The interior is simple but very functional with a set of Classic Instrument Ford gauges, a radio mounted under the dash, a Mullins steering column, and a cutout in the seat for the five-speed Hurst shifter. Sid Chavers did the white upholstery.

Following page: Roy Brizio takes his purple roadster out for a spin as often as he can, and he is not afraid to drive it cross country, even without a top. When the car was built in 1995, he drove it on the *Rod & Custom* Magazine America Cruize.

Mumford's 409 coupe is based on a '62 Chevy concept that includes the paint, the hubcaps, the interior, and, of course, the '62 Chevy 409-cubic-inch, 425-horsepower engine.

John Mumford's 409 Coupe

John Mumford is a big collector of old Fords, 1950s models, and old famous cars, such as the Sam Barris Mercury and the aforementioned *Ala Kart*. In addition, his shop shares a building with Brizio's. John tracked down a hard-to-find original 1932 Ford three-window and a NOS 1962 425-horsepower 409 Chevy engine for this project. John and Roy decided to build an early 1960s Chevy-style hot rod coupe.

Brizio's set up the original frame in the jig, boxed it, made new crossmembers in the center and rear, and installed a Model A front crossmember. The chassis was completed with a Halibrand quick-change 1940

Like all Brizio-built cars, the 409 coupe is nice underneath, with all components painted, chromed, or polished. The rear end is a '40 Ford with a polished Halibrand quick-change center section suspended with Aldan coil-over shocks and long ladder bars.

Sid Chavers stitched the 1962-style interior with the original Chevy material, and it even has an Impala speaker embedded in the seat. Roy Brizio and John Mumford came up with the '62 Chevy concept for the unchopped '32 three-window coupe.

rear end, long ladder bars, and Aldan coil-over shocks. Up front, Brizio's installed a 4-inch dropped I-beam axle with a split wishbone, '39 Ford–style spindles with Lincoln-type drum brakes, Buick drums, and aluminum finned covers. The driveline's 409-cubic-inch engine and four-speed Muncie transmission were connected and dropped into the chassis. Starting with stock 409 exhaust manifolds, Jack Stratton crafted an exhaust system that was dipped and polished before final installation.

The original three-window body and fenders were in good shape, and the car was assembled for fit and disassembled for finish work. Dan Laughlin Customs did the prep work and painted the coupe in 1962 Corvette fawn beige metallic. What really sets this concept build-up apart is the 1962 Chevy–style Sid Chavers interior with 1962 material, a crest in the seat, and a 1959 steering wheel. A Hurst shifter gives John a four on the floor, and the dash is filled with a set of old Stewart Warner gauges. When admirers look closely, they will notice a lot of extra details, including orange pinstriping that matches the rims with 'Vette hubcaps, a 25-louver hood, special headlights, and a louvered rear deck lid.

A lot of prep work was done at Dan Laughlin Customs, where the coupe was painted in '62 'Vette fawn beige metallic with orange Tommy the Greek–style pinstriping. The body looks just perfect, even around the louvers on the trunk lid.

Mumford found the mighty 409-cubic-inch Chevy NOS engine with 425 horsepower for sale. Under the big chromed air cleaner you will find two four-barrel carburetors, and the engine is painted in the original Chevy orange.

Ford made a lot of Fordor models, but not many are left today because nobody wanted them later on. Today, street rodders want all of the original '32 models.

Under the hood, a Roush-built, 402-cubic-inch, small-block Ford is backed up by a 700R transmission. The headers are part of the Brizio-made stainless-steel exhaust system, and Mustang rack-and-pinion steering hooks up to the Heidts Mustang-style front suspension.

Copper Four-Door

Seldom do any of the big-name street rod builders have a four-door 1930s Ford available for one of their customers, and it is even more rare to see a '32 Ford Fordor street rod. Although the Fordor was a cheaper model, it is difficult to find this body style in good shape today. Once it became difficult to find good '32 Fords, many street rodders turned to the alternative Fordor body style. Bob Nickum wanted a stock-looking but good-handling street rod, and he and Roy decided to build a rod with independent suspension in all four corners. The Brizio-built chassis uses a Heidts Mustang II–style front and rear suspension. For the driveline, a potent Roush-built, 402-cubic-inch, small-block Ford engine backs up a 700R transmission.

To get a softer ride in the sedan, Bob Nickum wanted independent suspension all around. Brizio's used Heidts suspension both front and rear, fully independent with Aldan coil-over shocks. Most of the components were chromed or polished stainless steel.

Bob's Fordor has a light-brown leather interior with matching brown carpet. The wood-grained dash with Auto Meter gauges and a wooden banjo steering wheel complete a classy look.

The vintage '32 Ford was in good shape, and Dan Laughlin completed the little repair work and straightening required. With the bodywork done, Dan Laughlin Customs prepped the body and painted it copper metallic. For the interior, Bob selected luxurious, classic, light brown leather upholstery that Sid Chavers stitched and paired with darker brown carpet. The dash was treated to old-style wood graining, a set of Auto Meter classic-style gauges, and a Stewart Warner–type nostalgic insert.

Nobody can walk past Gilbert "Bubba" Bugg's Pagan Gold full-fendered roadster without taking a closer look. The basic components are a Brookville replica steel body, small-block Ford engine, and five-speed transmission.

Andy Brizio is sometimes called "the Rodfather," and his son Roy builds some of the best street rods in the business. Steve Moal did the metalwork and the paint job in purple with white scallops on Andy's Brizio-built '32 three-window coupe. Cub Barnett built the 350/350 Chevy combination, and Sid Chavers crafted the gray leather interior.

For *Rod & Custom* magazine's 50th anniversary in 2003, this '32 roadster was built as a replica of the roadster that was featured on the first issue of the magazine. Brizio's built the roadster with a Brookville steel body, Ford small-block, five-speed transmission, and big billet steelie-type wheels.

Eric Clapton's roadster was painted PPG Baby Blue. It has a 351-cubic-inch Ford small-block under the hood and a Tremec five-speed transmission. The car was built to take his family on outings to the beach.

Roy built this project roadster with B&M and used a blower on the small-block engine. It has a Wescott fiberglass replica body on a Brizio frame, and Roy took it for long trips that earned his shop lots of ink in the magazines. This is Roy and his wife Jean Marie driving from Los Angeles to San Francisco. Andy Brizio is in the other car. *Steve Coonan*

Roy has covered many miles driving to events all over the United States with his roadsters. Steve Coonan snapped this shot while driving Roy's blue and red scalloped B&M/Brizio project roadster through the desert. *Steve Coonan*

In 1987, Roy Brizio and his team won the America's Most Beautiful Roadster award for the red Ferrari V-12-powered '32 roadster they built for James Ellis. Roy road-tests all Brizio cars before delivery, and this photo shows Roy driving the Ellis roadster. *Steve Coonan*

Dick Munz has several Brizio cars and a garage full of '32s, but he still wanted a basic Brizio roadster with a 351-cubic-inch Ford small-block and Tremec five-speed. The color combination of Sierra Gold and cream makes this fenderless roadster stand out.

5 '33 AND '34 FORDS

From a set of special Curbside frame rails, Roy Brizio and Jack Stratton made the frame and designed it to tilt the channeled body for a perfect stance. With a 1.5-inch shorter grille, a lower hood, and a 3-inch chopped top, the '33 is as good looking as the Tom Prufer *Cop Shop Coupe*, which was Jorge Zaragoza's inspiration for this project.

Dan Hall tuned an old drag race Enderle bug catcher fuel injector to perfection to get the maximum horsepower on the big rear tires. The hood was given an edge around the fuel injector, which forces the air into the bug catcher.

Blown Hemi '33 Coupe

When Jorge Zaragoza first got his license, he dreamed of owning a '33 or '34 Ford coupe. Years later, when he had the funds to buy or build his favorite hot rod, the car of choice was the classic Tom Prufer '34 *Cop Shop Coupe*. The original was not for sale, so Jorge and Roy Brizio made plans for a '33 three-window coupe based on the same concept.

The hot rod project became a mix of the Prufer coupe and Cliff Hanson's blown Hemi '34 coupe. Jorge's machine would have a 1960s race car feel, with a blown engine, big rear wheels, an interior roll cage, and more. Brizio located a good body for the project, and Bill Ganahl started on the metalwork. Jack Stratton built the coupe's frame with special

A Bruno Gianoli–built '58 Chrysler 392-cubic-inch engine was put together with a stock crank, 9:1 blower pistons, Hot Head aluminum heads, a rare extra-low Weiand manifold, and a 6-71 blower with an Enderle fuel injection system on top sticking through the hood.

Bill Ganahl and Jack Hagemann at Brizio's carefully chopped the top. The body was also channeled and tilted forward to get the right look, along with a tilted shorter grille and a lowered hood line.

frame rails from Curbside in Ohio, adding Model A frame horns to better fit the '33 grille. Like the *Cop Shop Coupe*, Jorge's coupe used a combination of '33-style and '32 rails in front of the firewall with an extra kick-up to lower the car. The channel of the body was tilted to give the car a better stance. Jack and Bill chopped the '33 coupe 3 inches in the Brizio shop, the rear wheelwells were tubbed for the big tires, and a new hood was crafted for the car's new look.

Bruno Gianoli built a '58 Chrysler 392-cubic-inch engine with 9:1 blower pistons, a hard-to-find low Weiand blower manifold, and a 6-71 blower with Enderle fuel injection. The engine was designed so that the injection system would stick out from the hood just enough to achieve the desired look. A manual five-speed transmission and Winters quick-change rear end completed the driveline. The front end has a 5-inch dropped tube axle with hairpins, a Posies leaf spring, and a set of disc brakes inside 4x15-inch Halibrand front wheels with skinny Michelin tires.

Darryl Hollenbeck completed the finish work and paint on the body. He painted the pieces in a red candy mix, and Art Himsl painted flames and pinstriping on the car before it was clear-coated. The interior was finished in black leather with a black mohair headliner, and the chromed roll cage was installed. Dan Hall fine-tuned the engine and fuel injection to get the car running smoothly while delivering as much of the coupe's 470 horsepower to the rear end as possible.

The interior of Jorge's coupe was done in black leather with red piping, and the fit around the chromed roll cage is as perfect as can be. The big tach on the Mullins steering column is part of the racing concept.

A Winters quick-change rear end was installed with ladder bars and a set of Aldan coil-over shocks. The deeper wheelwells on the rear end give the 11x16-inch Halibrand wheels and mega-big Hoosier tires some extra space. A panel under the body was made with punched louvers.

The Vicky body already had a 2-inch top chop when it was purchased for the project, and the Brizio team made the lines even better by shortening the grille 1.25 inches and modifying the hood top to drop the hood line.

A brand-new LS6 Corvette engine was detailed with chrome, polishing, and paint. A set of headers was used together with a Brizio-made exhaust system in steel, which was dipped. The firewall had to be cut to fit the big Corvette engine.

Mrs. Zaragoza's Driver

Married to a man who collects extraordinary hot rods, Paulette Zaragoza has picked up some interest too. She wanted her own street rod with a roll cage and other safety features so their daughter could ride in the car. Paulette was a Corvette fan before she met Jorge and owned a few over the years, but she researched her hot rod choice for some time. When she selected a '33 Ford Victoria, Jorge and Roy Brizio tried to talk her into a model that would be easier to acquire. But Paulette knew what she wanted and considered the '33 more elegant than the more common '32 Vicky. Eventually, Roy found a good '33 body with the top chopped 2 inches and the roof filled.

Sid Chavers did the interior in 1960s style with black-and-white leather and matching black carpet. The dash was modified with a Stewart Warner–style insert and a set of Auto Meter gauges. The steering column is a chromed Ididit with an early Corvette wheel on top.

The car was to be elegant and powerful with some 1960s style and a good ride. The team built a new chassis with boxed LRS frame rails and Heidts front and rear independent suspension. The firewall was modified to fit a brand-new Corvette LS6 engine, which was paired with an automatic transmission. The grille was lowered 1 1/4 inches to complement the hood line, and the hood was modified to fit the tilted and lowered line. Adding to the 1960s style, Wheel Vintiques replica Buick Skylark wire wheels were used in all four corners with thin whitewall tires.

Once Mickey Galloway finished the body and Brizio's completed the mechanicals, the car was dismantled for the finish work. Darryl Hollenbeck

It's easy to see the difference between the regular 1933/1934 Tudor sedan and this Victoria model with a shorter roof and sloping rear panel.

Darryl Hollenbeck gave the car the Pagan Gold paint job that makes the Vicky shine in 1960s candy style. The old Buick Skylark wire wheels with a 1-inch whitewall on the tires, the white top piece, and the running boards complete the 1960s look.

did the final prep work and painted the car in Pagan Gold candy. Once the parts were painted, the Brizio team put the body back on the rolling chassis for final assembly. Paulette directed Sid Chavers to stitch 1960s-style black-and-white leather upholstery. The Ididit steering column was topped with a Corvette steering wheel, and the dash was filled with an engine-turned insert and Auto Meter gauges. For the finishing touch, the running boards and a faux roof insert were covered in white upholstery.

Mel wanted to keep his phaeton looking stock but updated with modern high-performance equipment, even if he stuck with solid axles front and rear. Thanks to coil-over shocks, Panhard bars, and sway bars, the handling can be improved a lot.

Mel Taormino's 1934 Phaeton Touring

Once owned by Tex Smith and named a "Top 10" car in *Rod & Custom* magazine, this open four-seat '34 Ford has been in Mel Taormino's possession since 1969. Some time ago, Mel hired Brizio's to take the car apart and update it.

Starting with the bare frame in the frame jig, Jack Stratton installed parallel leaf springs with a 9-inch Ford rear end and a pair of tube shocks. The front end has a chromed Super Bell axle held in place by a pair of Pete & Jake's hairpins and a dropped leaf spring. Before the car was reassembled, the rolling chassis was completed with a 302 Ford small-block and a 700R automatic transmission. The headers, wiring, plumbing, and Brizio-made steel exhaust were done before the body was returned to the chassis. Guy

The interior of Mel Taormino's '34 touring is very impressive, and Sid Chavers stitched up the dark-green leather like a classic English sports car. With a wood-grained dash, wooden steering wheel, and original-style gauges, this interior has old-style class.

Brizio's cars are always tested before they are delivered to customers in order to work out any bugs and make sure there are no vibrations or other problems. Going down the freeway at traffic speed is a good test. The wind wings on the windshield keep some wind out of the interior.

A Ford in a Ford. Many might have picked a Chevy engine for a street rod like this, but Mel wanted a small-block Forc in combination with a 700R transmission. The high-performance, 302-cubic-inch engine was detailed with a polished manifold, Edelbrock valve covers, and a Mooneyes air cleaner.

Ruchenet worked the body to make it straight, and Lucky 7 Kustoms in Antioch prepped the body and fenders before painting them black with light green pinstriping. Light green detailing matches the Wheel Vintiques wheels and ties the car together.

To complement the classic, traditional look of the car, Sid Chavers stitched impressive, classic, dark green leather upholstery. The wood-grained dash was fitted with original-style gauges and a wooden-banjo steering wheel. Like most Brizio projects, the car has improved performance and handling but retains its original Ford styling.

With tan wire wheels, this red '34 roadster has a bit of English sports car feel. The body is a steel replica from Steve's Restoration, and the hood is an original Ford part.

Chuck Thornten's Full-fendered '34 Roadster

Chuck Thornten wanted a '34 roadster, which is often easier said than done—a '34 roadster for sale and in good shape is tough to find. Chuck and Roy Brizio decided instead to buy a replica steel body from Steve's Restoration in Oregon.

The team started with a Brizio-made frame with Curbside frame rails and a Kugel-made stainless-steel front suspension with coil-over shocks. The Kugel-made quick-change rear end with stainless-steel A-arms and dual coil-over shocks completes the suspension. A recessed firewall was modified

The '33/'34 Ford roadster has some nice lines with a chopped windshield and lowered top. The roadster has suicide doors, and the body is dropped down a bit over the frame as compared to a '32.

to fit the Ford Racing–built 392-cubic-inch small-block engine with four Weber carburetors. The powerful and sporty engine was connected to a five-speed Tremec transmission, and the car rolls on 16-inch replica wire wheels with Goodyear radial tires.

Dan Laughlin did the prep work and painted the car bright red, and Rory Pentecost added Tommy the Greek–style pinstriping along the beltline. Jim Vickery wired the car, and Sid Chavers stitched a light brown leather interior over a Chavers-modified seat with matching door panels and a cutout in the floor for the Hurst shifter. The beige carpet was edged in the same brown leather. Chuck's full-fendered red '34 roadster turned out just the way he wanted it.

Chuck wanted some extra horsepower under the hood, and Roy recommended a Ford Racing–built, bored-and-stroked, small-block, 392-cubic-inch Ford. Four Webers were installed to give the engine some extra sporty style. From the headers, the Brizio team built a stainless-steel exhaust system.

The interior is an important part of the finish work, and Sid Chavers stitched light brown leather upholstery with beige carpet. The steering column is an Ididit with tilt, and the dash got a set of Auto Meter gauges.

Plenty of traditions are built into this '34 roadster, even though it was built just a few years ago, because many of the ideas came from early hot rod magazines. *Steve Coonan*

George Poteet's '34 Roadster

When George Poteet contacted Roy Brizio about this project, he had a clear vision of a car that would be inspired by a couple of well-known hot rods—Jack Morgan's '34 roadster with a '37 truck grille, which was a cover car for *Hot Rod* magazine in 1951, and Jake Jacobs' yellow '34 coupe with a narrowed '37 truck grille.

At the time, no steel replica '34 roadster bodies were available, so the car was based on a fiberglass replica from Rat's Glass. The Brizio team created a new chassis with Curbside reproduction frame rails, a dropped Chassis Engineering I-beam axle, Pete & Jake's hairpins, and a 9-inch Ford rear end from Currie Enterprises with Pete & Jake's ladder bars, Aldan shocks, and a stock '40 Ford spring. For the drivetrain, a 351-cubic-inch Ford Racing engine with aluminum heads is backed up by a Borg-Warner T-5 five-speed transmission with a Hurst shifter.

With the body installed, Jim Hendricks narrowed a '37 truck front end and grille 3 inches before Jack Hagemann made the new hood for it. Camilleri's

The George Poteet fenderless '34 hot rod is a little unusual for a Brizio-built roadster. Big whitewalls on 16-inch steelies give the car a nostalgic style.

Steve Coonan

The unusual truck front end gives this roadster a lot of personality. The front suspension has a drilled I-beam axle, '39-style spindles, and a set of So-Cal disc brakes hidden in Buick drums. *Steve Coonan*

Auto Works took care of the finish work and painted the car in DuPont Concept Black before Rory Pentecost added Tommy the Greek–style pinstriping. To follow the old-style concept without sacrificing braking power, Brizio's used a set of So-Cal Speed Shop Wilwood disc brakes hidden in Buick drums up front and big 11-inch drums in the rear. George wanted 16-inch steel wheels with big whitewall tires and '46 Ford hubcaps with trim rings. An old-style interior means tuck and roll, and Sid Chavers did the work in red-and-white Naugahyde. The dash was given a So-Cal Speed Shop insert with Classic Instrument gauges and a Mullins steering column with a Bell-type sprint car steering wheel. The last work done at the interior shop was a white top over the chopped windshield. This little black roadster made a big impression on the hot rod world.

Without sunshine, this '33 cabriolet might look black, but here it is easy to see that it is in fact deep purple. The '33 rolls on a set of 15-inch front and 16-inch rear '35 Ford wire wheel replicas with Goodyear radial tires.

Deep-Purple '33 Cabriolet

It's easy to see that Jorge Zaragoza likes '33/'34 Fords. This stylish, Brizio-built, full-fendered, purple cabriolet is diplomatic but very well built. The difference between a '33 roadster and a '33 cabriolet is that the latter has door glass and the windshield posts are part of the body.

This project started with a good original body and a Brizio-made frame. Jorge wanted front and rear independent suspension, so Brizio's installed a stainless-steel Kugel front suspension and a Kugel quick-change independent rear end with stainless-steel arms, billet uprights, and front and rear coil-over shocks. Extra work with the chassis made this 75-year-old Ford drive like a sports car,

A look underneath the '33 reveals an impressive Kugel quick-change rear end with inboard Willwood disc brakes. The suspension consists of polished stainless-steel A-arms, billet aluminum uprights, and a dual set of coil-over shocks.

The interior is pure luxury, and a Glide seat was covered with the red leather upholstery. Like most of Jorge Zaragoza's street rods, the '33 has a Tremec five-speed transmission with a Hurst shifter on the floor.

Ford Racing bored and stroked this small-block Ford engine to 392 cubic inches. The brand-new crate engine was delivered to Brizio's, and the Brizio team detailed the engine by polishing the aluminum, chrome, and stainless steel.

You have to know a lot about '33/'34 Fords to see and realize that Brizio's shortened the grille 1.5 inches to perfect the hood line. Because of that modification, Jack Hagemann made a completely new aluminum hood. The modified hood line, chopped top, and bobbed rear fenders give the car a great look.

especially with sway bars and a set of big disc brakes all around. A Ford Racing–built 392-cubic-inch small-block Ford engine with plenty of nice detailing works with a T-5 five-speed transmission controlled by a Hurst shifter.

The original body was left much the same as it had come from the factory, with minor changes, including hidden door hinges, bobbed rear fenders, and chopped windshield posts. The grille was cut down 1 1/2 inches for a better hood line, and Jack Hagemann crafted the hood in aluminum. Vintage Color Studios did the finish work and painted the car in deep purple paint. Sid Chavers covered a Glide seat with red leather upholstery, and the seat and door panels were upholstered to match. Chavers put in coordinating carpet with red leather edges and crafted a lowered cabriolet top. A full set of Auto Meter gauges fills the dash, which includes small outlets for air conditioning. The Ididit tilt-type steering column connects to a banjo steering wheel.

6 '35 TO '40 FORDS

Many rodders think that a '40 coupe needs bumpers to look right, but Clapton's coupe looks great even without them. The scoop sticking through the hood hints at the potent engine under the hood.

Out on the road, the coupe comes to its right and it is clear how well the car is designed. Rolling down the highway, the coupe's polished aluminum wheels and scoop shine in the San Francisco sunlight.

Eric Clapton's 1940 Ford Rock-and-Roll Coupe

Eric Clapton's coupe is such a dark shade of blue that it looks black, and the glossy old 1940 Ford provides the rock star with some welcomed anonymity. When a street rod this beautiful drives past, observers are too focused on the car to notice the living legend behind the wheel.

This street rod was Eric's first, and he was very much involved with the details. For a clean look, Brizio's cut off the drop rails on top and removed the door handles, as well as the handles and chrome from the hood and trunk. Because Eric and Roy decided to go without bumpers, the bumper bracket holes were filled in. The headlights were fitted flush in the fenders, and a stainless-steel grille was handmade for the front end. The Brizio team made a big taillight that is flush with the rear body panel and installed a third brake light under the rear window.

Clapton had one of his favorite expressions engraved in the right valve cover: "Went down to the crossroads fell down on my knees," by Robert Johnson.

Roy Brizio Street Rods got started with Eric Clapton's coupe about a month after the plan to build the car was done. Roy is actively involved in all the cars being built at his shop and is pictured here with the '40 chassis in the jig.

Camilleri's Auto Works painted the body and fenders, and the end result was as perfect as a mirror, which is a sign of high-quality bodywork. The independent front suspension and coil-over shocks in the rear make for a perfect ride. Sid Chavers stitched the interior in beige leather over a Glide Engineering seat. For a bit of race car feel, there is a chromed roll bar behind the seat. A complete set of Auto Meter gauges are sunken into the dash, and the steering wheel is from Billet Specialties. The Hurst shifter on the floor is bolted to a six-speed Richmond transmission, which gives the Ford even more of a sports car feel. With 687 horsepower from the 540-cubic-inch blown big-block, the car pulls away strong even at lower rpms. Eric was so pleased with his first street rod that he plans to have Roy Brizio build him another one in the near future.

The powerful engine in Clapton's '40 is based on a Merlin block with a Lunati crank, Brodix aluminum heads, and a BDS 8-71 blower on top, plus two polished Edelbrock carburetors. Jack Stratton built the 3-inch exhaust.

After some time on the road, Clapton asked Roy to change the coupe a little bit, swapping the big blown engine for an easier small-block and filling in the hole in the hood. The coupe also has bumpers now, front and rear.

The big lift—Paul Cammileri (in the white T-shirt) and his crew drop the body onto the chassis.

A 540-cubic-inch big-block Chevy is impressive, but Clapton wanted a blower and a scoop sticking through the hood, too. The fully polished Chevy engine was built by John and Bruno Gianoli with only the best in components. On the dyno run, Clapton had a chance to see his engine produce 687 horsepower.

Out in the sun, paint will show every little uneven spot, but Camilleri's made sure the bodywork was perfectly straight before it was painted. The body was cleaned of all the unnecessary material, including drip rails on the roof, handles, and holes for the bumper brackets. Check out how well the taillight melts into the body lines.

The original dash was filled before a set of Auto Meter gauges was placed in it. An Ididit steering column was installed with a Billet Specialties steering wheel. The Hurst shifter connects to a six-speed Richmond transmission.

With the body dropped over the frame, the rumble seat lid was changed to a regular trunk lid. The rear suspension and tank needed more space under the lowered body. Jack Hagemann made a new and smaller aluminium bumper with a pair of handmade taillights mounted inside.

New-Old Custom Rod

Many of the old customs have a history, which also means people are willing to pay more money for them, and more people are looking for them. After years of searching, Bob Rosenthal gave up on finding an old custom and focused on building his dream car from a '39/'40 Ford instead. Rebuilding allows suspension updates with modern components and more creature comforts while maintaining the car's old-style look, and Bob hired Roy Brizio to make it happen.

When the car was delivered to Roy Brizio Street Rods, the crew set up a new chassis in the jig with a complete Kugel stainless-steel front and rear suspension. The independent setups, front and rear, are hidden under the lowered bodywork. The independent suspension gives the car a smooth ride, and it reacts more like a modern car. Bob wanted to keep the old-style look overall and selected a set of 16- and 17-inch Wheel Vintiques billet wheels with a steelies design. For some modern horsepower under the hood, Bob chose an LS1 Corvette fuel-injected small-block Chevy and a late-model four-speed automatic transmission with a Lokar shifter. The body was channelled 4 inches for an old-fashioned custom look, and the chassis and body were much lower than original. Metal man Mickey Galloway built a new floor and firewall, and modified

The lines of the '39 convertible are smoothed out, thanks to the channelled body and a slight chop of the top and windshield. The wheelwells were opened up to fit a new set of wheels and larger-diameter rims.

and fitted the front inner fenders. He also sectioned the original hood to flatten it and chopped the windshield a few inches. With the metalwork done, Lucky 7 Kustoms did the finish work and painted the car a deep cherry red, which Bob had selected long before the car was ready to paint. A black base makes the red darker and more mysterious.

Back at the Brizio shop, the car was put back together and detailed. With wiring, stereo, and air conditioning installed, the car was sent to Sid Chavers. Sid made the seats and covered the aluminum dash in the same tan leather as the seats. The dash was treated to a full set of Auto Meter gauges and special chrome trim. The Ididit steering column hooks up to Mustang-style rack-and-pinion steering, which is controlled by a Juliano's wooden steering wheel. Carpet matching the upholstery was edged in tan leather, and the new tan convertible top complements the interior and exterior colors. It was an exciting day when Bob got to test-drive his new-old '39 custom rod, and performance Wilwood disc brakes in all four corners kept the fun in check.

The fuel-injected LS1 Corvette engine was detailed with exterior paint and polished parts before it was lifted into the engine bay. The fresh horsepower and smooth power curve give Bob more power than he needs.

Sid Chavers crafted the interior in tan leather and made the special seats and the console between them. He covered a custom-made dash by Jack Hageman with the same tan leather.

The new Brizio chassis has independent suspension all around, and the Wheel Vintiques wheels with Goodyear 225/16 and 275/17 tires give the coupe a perfect stance.

Jorge Zaragoza's Golden '36 Coupe

Jorge Zaragoza bought a near-stock '36 Ford three-window coupe—one of his favorite models—in fairly good, rust-free condition in Texas. Once it arrived at Brizio's in San Francisco, the car was dismantled. Jack Stratton built a new chassis with Curbside frame rails, a Heidts front suspension, and a complete Heidts independent rear end with Aldan coil-over shocks. Meanwhile, the coupe was sent to Marcel's Custom Metal for a 3-inch top chop. When the blasted and chopped body was matched up with the new chassis, Brizio's attached a shorter grille from Grille Art, Jack Hagemann made a new aluminum hood with a lower line, and Mickey Galloway finished the bodywork.

For the drivetrain, a Roush-built 427-cubic-inch Ford engine is used with a Tremec five-speed transmission.

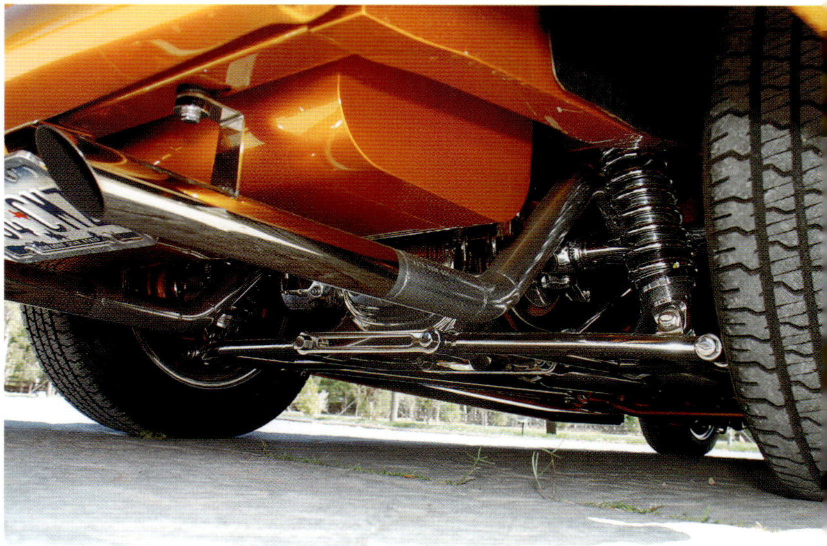

The rear end is a fully independent Heidts setup with Aldan coil-over shocks and stainless-steel A-arms. Big Wilwood disc brakes all around make it easier for Jorge to stop the coupe after he hammers the throttle pedal.

To improve the lines on the front end, a 1.5-inch shorter stainless-steel Grille Art grille was installed, and Jack Hagemann made a new aluminium hood with louvered side panels.

The powerful and good-looking engine was detailed with Roush aluminum heads, a polished Edelbrock manifold, a Demon 750 carburetor, Edelbrock valve covers, and a Mooneyes air cleaner. With all the mechanical parts together and the car pre-assembled, the team tore it down to have everything chromed and painted.

Darryl Hollenbeck took care of the finish work and prep work, and he painted the car in a glowing custom-mixed gold accented by Rory Pentecost pin-striping. Sid Chavers finished the interior in white leather, purple carpet, and black outlines. A 1940 dash was filled and modified for a set of Auto Meter gauges, and a new Ididit steering column was topped with a smaller 1940 Ford steering wheel. The golden coupe rolls on 16- and 17-inch Wheel Vintiques billet genies with Goodyear Eagle 225/275 tires. The coupe is much more than a looker; it's a perfect cruiser too.

A Bitchin firewall gives the engine some more space, and the 515-horsepower 427-cubic-inch Roush engine is dressed up with aluminum Roush heads, a polished Edelbrock manifold and valve covers, and a 750-cfm Demon carburetor under the Mooneyes air cleaner.

Inside, the 1940 dash was modified and filled with a set of Auto Meter gauges. Air conditioning outlets peek out from under the dash, and an Ididit steering column is paired with a smaller '40 Ford steering wheel. Sid Chavers stitched the white interior with purple carpeting.

Marcel's chopped the top 3 inches, and Jack Hagemann made a new lower hood, giving the coupe much cleaner lines. The brand-new Brizio chassis has independent Heidts suspension in all four corners with Aldan coil-over shocks.

The long lines of the '36 coupe were improved by a 3-inch top chop, fender skirts, and a dropped chassis. Custom bumpers front and rear are '41 Lincoln, and the '41 Hudson taillights sit beside a Frenched 1947 license plate.

The *Calori Coupe*

Jack Calori was a cop in Lynwood, California, and later in Long Beach. He built a '29 roadster with upswept pipes and a tuned flathead that he ran out on the lakebeds. One day in 1947, Jack ran across a '36 Ford coupe for sale that he thought would be a good tow car for the roadster. He bought the coupe and showed it to his friend Herb Reneau, who had a body shop and did modifications on cars for some hot rodders in the area. Herb talked Jack into modifying the coupe with a 3-inch top chop, a LaSalle grille on the front end, low-mounted '40 Chevy headlights, and '41 Hudson taillights next to a Frenched license plate. To lower the '36, Herb used the back half of another frame to make a big kick-up in the rear, which dropped the car even with the stock spring in the rear. The front was lowered with a dropped I-beam axle. Bubble fender skirts were installed in

There was enough of the original custom interior left for Sid Chavers to stitch a new one in the same type of material and style. The dash has a full set of Stewart Warner gauges, and the floor has a dark-red matching carpet.

the rear, and '41 Lincoln bumpers were added front and rear. The car was painted in black lacquer, and a Long Beach upholstery shop crafted the red interior. With its new look, the car graced the cover of *Hot Rod* magazine in November 1949. Some years later, Calori sold the car to a dealer as down payment on a '50 Mercury. He didn't see the '36 Ford coupe again for 50 years.

The old coupe was found in Spokane, Washington, and collector Don Orosco bought and stored it. Later, Jorge Zaragoza jumped on a chance to buy it. He hired Roy Brizio and Bill Ganahl to restore the car. Bill did

With Bill Ganahl as the project manager, Roy Brizio and his team restored this famous custom coupe to better shape than it was in during the late 1940s. Herb Reneau modified the coupe for Jack Calori in 1947 by chopping the top, adding a LaSalle grille, and dropping the chassis front and rear.

a lot of research before they started taking the car apart, and they even contacted Jack Calori to serve as a consultant on the project with the end goal of restoring the car to its 1949 appearance. One fender needed to be replaced, but the rest of the bodywork was in good shape. The chassis was restored, and Bruno Gianoli rebuilt the original flathead engine. The team tried to match every little detail, including perfect aftermarket hubcaps. Darryl Hollenbeck painted the car in a PPG black, and Sid Chavers matched and re-created the old upholstery's color and style. When the car was restored, Jorge Zaragoza took it to Pebble Beach, where he wanted Roy Brizio and Jack Calori to drive it up on the podium to receive the first-place award. And that is just what they did.

Originally, Jack Calori had the coupe engine in his '29 lakester, but he dropped his good engine in later. It was rebuilt with most of the original parts, a pair of Eddie Meyer heads from Don Orosco, and a manifold with two Ford carburetors.

To make the '30s Ford look even better than it did when it left the factory, Roy dropped the hood line just a little bit. On Peter Prescott's phaeton, the grille was shortened 1 1/2 inches, and Jack Hagemann made a new three-piece hood for it.

Peter Prescott's 1935 Phaeton

Early on, many hot rodders did not like the '35 Ford grille and front end, and some cars were modified with '36 Ford front ends instead. Today, most '35s retain their original grilles. One example is Peter Prescott's '35 phaeton in shiny blue candy. Peter's project started with a fairly good original body and a new Brizio-built chassis. Roy and Peter worked out the details for the car, including a Heidts Mustang II–style front suspension with polished stainless-steel A-arms with coil-over shocks. For the rear, a 9-inch Ford rearend was installed with Chassis Engineering leaf springs.

Peter wanted some extra horsepower in the phaeton, and suggested a hot, Roush-built 427-cubic-inch, small-block Ford engine with Weber-style fuel injection. The engine was backed up by a Tremec five-speed transmission, which makes the phaeton awfully fun to drive. With the body on the chassis, Jim Hendricks cut down the grille 1 1/2 inches, and

The '35 phaeton is bigger than the '33–'34, and the '34's suicide front doors were changed to regular doors on the '35 model. Vintage Color Studios completed the prep work and the beautiful blue candy paint job.

Jack Hagemann made a custom three-piece hood. A pair of smooth 1940 Ford bumpers completed the look. Darryl Hollenbeck at Vintage Color Studios did the paint and bodywork. To make the candy paint look good, Darryl's prep work had to be absolutely flawless, and he took the time to make it perfect. Inside, a modified 1940 dash was paired with a small Juliano's 1940 Ford steering wheel atop an Ididit steering column. As usual, Sid Chavers stitched the black-and-white leather upholstery and made a white top to fit the low and tilted DuVall-style windshield. The phaeton rolls on a set of Coker whitewalls on chromed Wheel Vintiques reversed 1950 Mercury rims with 1950 hubcaps. It was a crowd favorite at the 2008 Grand National Roadster Show.

A 1940 dash was modified to fit in the 1935, and an extra shelf was created under the dash to mount the radio. The Hurst shifter on the floor is hooked up to a Tremec five-speed transmission.

The powerful Roush-built small-block Ford engine has big-block volume with 427 cubic inches and an impressive fuel-injection system that looks like a Weber setup. As much of the engine was polished as possible, including the Edelbrock valve covers, alternator, and pulleys. The firewall was cut to make room for the engine.

A '37 Ford roadster is not easy to track down, but Paul Bonderson found one and hired Brizio's to restore it. Camilleri's Auto Works painted the clean body in a stock-type dark blue.

A Rare '37 Roadster

Hot rodders have long discussed which was the last real roadster: the '36 Ford or '37 Ford. The '36 has no glass in the doors and a separate windshield, and the '37 has no glass in the doors but a windshield frame that is part of the body. As a somewhat-official ruling on the longtime discussion, the Grand National Roadster Show recognizes the '37 Ford in its roadster class.

The '37 roadster is very hard to find because Ford only made 1,250 at its factory. Paul Bonderson had the good fortune of buying a '37 and hired Roy Brizio to restore and update the car while keeping most of its stock shape. Not all Brizio cars are modern nostalgia-style street rods; some are resto rods, and Paul's '37 is a good example. Paul's wish list included a blown flathead engine, and Brizio's ordered such an engine from H&H with a stroker Scat crank and Navarro aluminum heads and manifold with a new replica S.Co.T. blower with two 97 Stromberg carburetors. While H&H was building the flathead, the Brizio team dismantled the car and rebuilt the chassis. The

Paul Bonderson's very impressive engine is an H&H-built, bored-and-stroked flathead with a S.Co.T. blower, two 97 Stromberg carburetors, and a set of polished Navarro aluminum heads.

Sid Chavers crafted the simple and clean interior in beige leather. The dash was painted with wood graining, and the original instrumentation was retained. The floor shifter connects to a '39 Ford three-speed transmission with Lincoln gears.

This beautiful old Ford appears to be a convertible, but an up-close view reveals that its doors have no glass—it is the last of the Ford roadsters.

front axle was replaced with a dropped I-beam, and the front suspension was updated with a sway bar. The rear suspension was dropped with a re-arched spring. When the chassis was done, the new engine and rebuilt '39 transmission with Lincoln gears were lifted into the frame. Then the good-looking roadster body was dropped back on the chassis. The wood-grained dash and inside windshield frames give the interior an old-fashioned style, and the instrumentation remains stock, down to the banjo steering wheel.

Paul chose a dark blue for the exterior paint, and Camilleri's Auto Works did the finish, prep, and paint work. Brizio's used old-style wires in keeping with the resto-rod concept. The interior was finished in beige leather in simple, old-style patterns. Few would suspect Paul's unassuming vintage roadster is hiding a blown, bored, and stroked flathead under the hood.

The convertible's body line was much lower after the body was channelled over the frame, and the windshield frame and top were chopped. Door corners were rounded, and the bottom of each fender was modified.

Mumford Custom '38 Convertible

John Mumford likes channeled old customs, and that's just what he had in mind when he bought a '38 Ford convertible in good shape.

John sent the car to Brizio's, where the body was lifted off the chassis. The team stripped and sandblasted the frame before it was set up in the frame jig. A new Heidts front crossmember was welded in place for the Mustang II–style front suspension with stainless-steel A-arms and coil-over shocks. For the rear suspension, a 9-inch Ford rear end was fitted with a set of Chassis Engineering leaf springs for a softer ride. The old flathead was updated to a brand-new, 385-horsepower small-block Chevy with an Edelbrock carburetor and plenty of detailing. Once the five-speed Tremec transmission was bolted on the engine, the combo was dropped into the chassis. The body was then fitted back on, and the floor was cut out because it would be channelled a few inches over the frame. The new floor has a tunnel over the transmission, and a new firewall was fitted behind the engine. With the body a little lower on the chassis, the windshield was chopped a few inches to improve the car's overall look. The '38 Ford's high rounded hood top was sectioned down, the fender bottoms were modified, and the running boards were removed. The Brizio team filled the bumper bracket holes and installed a pair of '39 Ford taillights.

Lucky 7 Kustoms did the finish and prep work, and Marcos Garcia painted the car a dark green candy. On the inside, Sid Chavers stitched a classy light brown leather interior. The dash was left stock and the gauges restored, while the original steering column was modified to hook up to the Mustang II rack-and-pinion steering. The convertible rolls on a set of Wheel Vintiques old-style wheels with hubcaps and Goodyear radial tires.

The body was channelled over the frame a few inches, and the hood top was sectioned. The brand-new, 383-horsepower Chevy crate engine with aluminum heads, manifold, and carburetor was detailed with Mooneyes valve covers and an air cleaner.

Mumford's '38 Ford has an independent Heidts front suspension and a 9-inch rear end with Chassis Engineering leaf springs. The hood top was sectioned to flatten it and achieve a lower body line.

The classic interior was outfitted with restored original gauges and a modified original steering column with a banjo-style Ford steering wheel. Seat and door panels were upholstered in light-brown leather, and the green carpet matches the green exterior.

Brizio's built this '36 coupe with a 350/350 Chevy engine/trans combination. Jorge Zaragoza of El Paso now owns the good driver.

Dick Munz is a close friend and longtime customer of Roy Brizio Street Rods. Brizio's built this '36 roadster with a small-block Ford 351-cubic-inch engine and a five-speed transmission. Here the car is on its maiden voyage, heading from San Francisco to Madison, Wisconsin.

7 '50S TO '60S

Roy built this '55 Chevy to drive it to hot rod events all over the country.

The clean-looking '55 Chevy sits on a new Art Morrison chassis with an independent front suspension, a 9-inch Ford rear end, and Strange Engineering coil-over shocks all around.

The Golden Shoebox

People often associate Roy Brizio's name with the good-looking street rod drivers his company builds. And while '32 Fords dominate the shop, his team has also built and restored a long list of 1950s cars over the last few years. When Roy got an offer to buy a good '55 Chevy 210—the two-doorpost model—he couldn't pass it up. Roy's crew started on the '55 by dismantling it down to the bare body, which was then media-blasted with nutshells (which penetrate to the bare metal without affecting the body's shape). At one time, a racer had done a poor job widening the rear fenders to make space for slicks, so new pieces were ordered, and the rear fenders were returned

Roy and his crew are really good at detailing, which is easy to see under the hood. The brand-new Edelbrock 350 crate engine was assembled with aluminum heads, an Edelbrock cam, and a manifold with two Edelbrock 600-cfm carburetors. Sanderson made the headers, and the stainless-steel exhaust was finished off with Edelbrock mufflers.

The '55 sits low like a modern street custom, but Roy also used a lot of semi-custom styling for his two-door Chevy.

to stock appearance. Fortunately for Roy, the floor was in good condition and free of rust, making the restoration process much easier.

While the body was being fixed, Roy ordered a new chassis from Art Morrison with a Morrison front suspension and tube A-arms, Strange Engineering coil-over shocks, and 2-inch dropped spindles. The rear suspension has a Morrison four-bar setup with a 9-inch Ford rear end, Strange Engineering coil-over shocks, and Wilwood disc brakes.

Ordering the new chassis instead of building it saved Brizio's many hours of labor. Roy wanted a modern yet old-style mild custom, so the crew filled the emblem and some chrome trim holes on the body, filled the bumpers and installed them tighter to the body, louvered the hood, and installed a custom grille. For the driveline, Roy hired Edelbrock to build a crate engine based on a Chevy 350 ZZ4 with Edelbrock heads, camshaft, manifold with dual quads, and an MSD distributor. Roy called on his next-door neighbors at Sanderson to make a set of headers for the Chevy. Behind the potent powerplant, a five-speed Tremec transmission with overdrive was bolted on to work with the 3.70:1 rear end.

With the car pre-assembled, it was time to prepare it for paint. Roy contemplated the color for some time and test-painted panels with different shades of orange and gold. He selected a special mix of DuPont Pagan Gold that makes the car shine. "I wanted the car to be seen," Roy says with a smile.

Sid Chavers stitched the black-and-white Naugahyde upholstery and headliner in an old-style tuck and roll. A 1950s Corvette steering wheel tops an Ididit steering column.

To finish the car, he asked Sid Chavers to stitch a new interior in old-style black and white. The dash was given a set of Classic Instrument gauges, and the Ididit steering column has a '62 Corvette wheel on top. To get the car rolling, Roy had his friends at Billet Specialties make a special set of 17- and 18-inch wheels, which were shoed by a set of Diamond Back whitewalls. Roy showed off the Chevy at the San Francisco Rod, Custom, and Motorcycle Show, and the Grand National Roadster Show, and it was a great success. "I like to show my customers that we build nice fifties cars too," Roy says.

Mumford and Brizio brought the Barris custom out to Paso Robles in 2007 and showed it before the upholstery was done. Sam Barris originally did a full-fadeaway side line from the top of the front fender to the end of the rear fender.

The Sam Barris Mercury has stock dash instrumentation. Two Appleton spotlights were added in traditional custom style. *Steve Coonan*

Sam Barris '49 Mercury

Roy Brizio's next-door neighbor, John Mumford, was meeting with show promoter Bob Laravee when he asked Bob about the old Sam Barris Merc.

"Do you know what happened to it?" John asked.

"Better than that, I know where it is right now," Bob said.

Laravee's friend Tommy Lee had the car in New Jersey. And it just so happened that it was available. And so 47 years *to the day* after Tommy bought it, Roy and John showed up to buy it from him. The car had had some experiences over the decades—it had even been stolen once. Before Tommy had bought it, a few other owners had done some modifications, and many details—including the fender skirts, door panels, hubcaps, and sun visors—were gone. It had

This photo shows the Sam Barris Mercury when Roy Brizio and its new owner John Mumford first saw it in Tommy Lee's garage on the East Coast. Mumford bought the car and transported it across the country to San Francisco.

undergone a few engine swaps, too, at one time being powered by an Olds 303, then a Buick Nailhead, and a few other engines. But Tommy Lee had saved the car from rusting through the years, and he had the original frame.

After John Mumford bought the car, Brizio's set to work restoring it to its original Sam Barris–built condition. Bill Ganahl was the project manager and researched the car to make sure every detail was correct. The Merc had been brand new when Sam Barris chopped the top 4 inches in the front and 5 1/2 in the rear. The late custom legend also extended the fenderline into a full fadeaway and used '48 Buick chrome trim. The headlights were Frenched, the door handles shaved off and holes filled in, the hood nosed, and the deck lid shaved. The grille opening was modified for a '51 grille. Many said Sam Barris was one of the best sheetmetal guys they had ever seen.

For the original build, Barris left the chassis and drivetrain near stock, with the 255-cubic-inch flathead and overdrive three-speed transmission intact. The rear springs were re-arched to lower the car, and the frame was "C'd" to give it some more travel. The front springs were cut one coil to make

Continued on page 155

Following spreads: Sam Barris bought his '49 Mercury new as soon as the cars were delivered to the dealers, and he started to think about how to chop it and make other modifications. The car was chopped and modified at the Barris shop in Lynwood, California. *Steve Coonan*

60 D 631

60 D 631

After it was restored by Roy Brizio Street Rods, the Sam Barris Mercury was displayed at the Grand National Roadster Show in January 2008. The Merc is considered *the custom* that set the trends for the 1949–1951 custom Mercurys over the past 50 years.

To complete the restoration, Ron Mangus crafted a new interior for the Merc that was very similar to the original Barris interior. The white vinyl has green piping with pleats in the seat and the door panels. *Steve Coonan*

Continued from page 149

the Merc sit right. Bill Ganahl restored the original frame and Bruno Gianoli restored the 255 flathead. Then it was time for the restored body to be moved over to the chassis.

Darryl Hollenbeck at Vintage Color Studios painted the car in the original dark green metallic, matching the paint to a part of the dash that still had the original paint. It was a little harder to re-create the original custom interior because nothing was left of it, and no good pictures could be found. Ron Mangus stitched the interior in white Naugahyde with green piping, and the headliner was done with green velour and white piping. The beautiful finished product is a sight to behold, and it can't help but make any hot rodder imagine the reaction when Sam Barris first unveiled it all those years ago. No one had seen a chopped '49 Merc before this one.

In dirty work pants, Sam Barris poses with his just-finished '49 Mercury. Sam was the first to figure out how to chop the Merc, cutting 4 inches from the front pillars and 5.5 inches from the rear.

The Brizio team set up the new Art Morrison chassis on the chassis table with the front Mustang II–type suspension in place and the 9-inch Ford rear end with four-link. With the chassis at ride height, the body was dropped on.

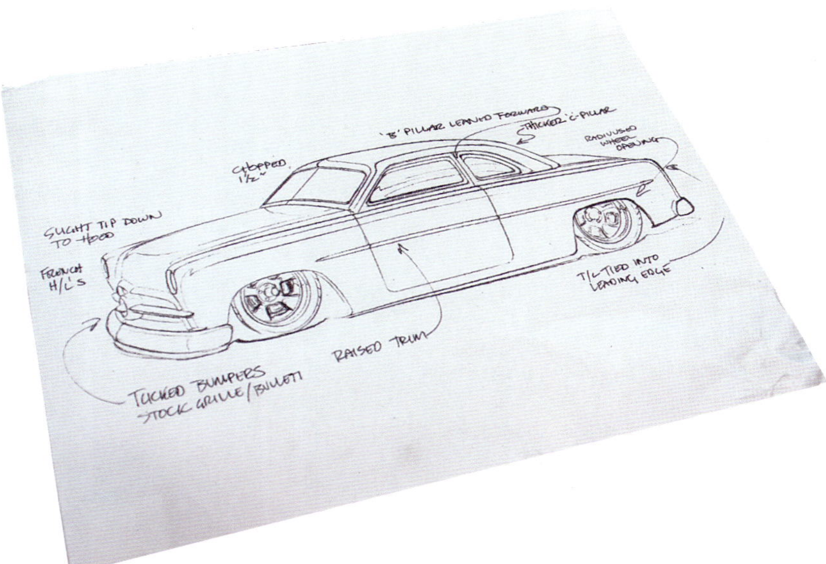

Thom Taylor's rendering of the '49 Ford coupe points out all the modifications.

New Project: Eric Clapton's '49 Ford Coupe

One of two new projects in the works for Eric Clapton is a '49 Ford coupe. It will be a mild custom with a slight top chop, Frenched headlights, and tunneled taillights. With a new, more modern front suspension setup and a new rear suspension, Roy and Eric figured that a new frame with all the suspension brackets done would be a good idea and would save time.

The updated chassis is based on an Art Morrison frame with a Mustang II–style front suspension with coil-over shocks front and rear, hooked to the 9-inch Ford rear end with a four-link. Among many other tasks, Bill Ganahl and Jack Stratton are doing the metalwork on the '49. The body had some rust when Eric bought it, and some panels had to be replaced, including the bottoms of both rear fenders, the bottom of the right door, and pieces on the front

Billy Ganahl is working on the inside of the deck lid. It is easy to see the replaced panels on the body, but the 1.5-inch top chop is more subtle.

With most of the body and sheetmetal work done, the Roush-built, 402-cubic-inch, small-block Ford engine and Tremec transmission were fitted in the car. The inner fenders were done with space for headers and other components.

Most of the bodywork has been done on the '49 with Frenched headlights and a chopped top, which can be seen with the cuts and welding on the A-pillars. The hood line will also be lowered a bit in the center.

Both rear fenders got new panels in the bottom, and a new panel was installed under the deck lid in the rear. The taillights were tunneled a bit, and holes were filled in the trunk lid.

fenders. The biggest modification done to the body is the 1 1/2-inch top chop, which took several hours to do, even though most people will not even realize that the finished car has been chopped. Other sheet-metal modifications will include dropping the hood line and opening the rear wheelwells, as shown in the rendering. The new engine is a Roush-built, 402-cubic-inch, small-block Ford with a five-speed Tremec transmission behind it. A new firewall and inner fenders will clean up the engine bay. But there will be something very unusual about Eric Clapton's '49: it will have the steering on the right side of the car because Eric is planning to take this Ford to England.

Rock legend Neil Young came by the Brizio shop with a problem on his Cadillac. That visit led to hiring Brizio's for a mechanical restoration of his '53 Buick convertible.

INDEX